D. Todt P. Goedeking D. Symmes (Eds.)

Primate Vocal Communication

With 76 Figures

Springer-Verlag Berlin Heidelberg GmbH

Professor Dr. DIETMAR TODT
Dr. PHILIPP GOEDEKING
Institut für Verhaltensbiologie
Freie Universität Berlin
Haderslebenerstr. 9
1000 Berlin 41, FRG

Dr. DAVID SYMMES
National Institute of Child Health
and Human Development
Laboratory of Comparative Ethology
Bethesda, MD 20892, USA

ISBN 978-3-642-73771-8 ISBN 978-3-642-73769-5 (eBook)
DOI 10.1007/978-3-642-73769-5

Library of Congress Cataloging-in-Publication Data Primate vocal communication / [edited by] D. Todt, P. Goedeking, D. Symmes. p.cm. Includes index. ISBN 0-387-19355-3 (U.S.) 1. Primates – Behavior. 2. Animal communication. I. Todt, D. (Dietmar), 1935– . II. Goedeking, P. (Philipp), 1956– . III. Symmes, D. (David), 1929– . QL737.P9P6754 1988 599.8'0459–dc19 88-19086 CIP

Softcover reprint of the hardcover 1st edition 1988

Typesetting, printing and binding: Brühlsche Universitätsdruckerei, Giessen
2131/3130-543210 – Printed on acid-free paper

Preface

The scope of this volume is to reflect current topics, approaches, and methodologies in the study of primate vocal communication. Most chapters are based on contributions that were presented orally at a symposium entitled "New approaches towards an understanding of primate vocal communication" held at the Freie Universität Berlin in August, 1986. The editors have solicited comments and in some cases contributions from investigators working in this area who did not attend the Berlin symposium, in an effort to present as comprehensive a perspective as possible.

It may be surmised that progress in this field is rapid from the decreasing interval between related international meetings and between volumes published on primate communication. This book follows more or less directly upon that edited by C.T. Snowdon, C.R. Brown, and M.R. Petersen which was based upon a symposium held in 1980 at Parma. Many of the same contributors appear in this volume, suggesting both healthy continuity, and possibly less admirable, a difficulty in expanding the field, no doubt attributable to some extent to the worldwide shortage of research funds. At any rate, since the interval between relevant volumes (15 years preceding the Snowdon, Brown, and Petersen offering, and about 7 preceding this effort) is declining, we may expect more students and perhaps some non-primate-oriented ethologists to be recruited.

We find that several trends have emerged in our field. First, older issues (controversies may be an acceptable characterization) have been swept along by a pragmatic, multidimensional approach, which tries to avoid the pitfalls of call-type classification and reduction to simple sender-recipient models, and has instead conceived of vocal systems as part of social organization involving many speakers and many listeners of different age-sex classes. Thus there is more interest in groups and large audiences. Clearly we now find greater emphasis on the way calls are used in narrow contexts and on subtle variation in acoustic detail than on the number of calls in a specific repertoire. The significance of acquired communication skills in the great apes, which was "in vogue" at the time of the previously cited contribution, seems less vital today. As predicted then by our colleagues, the ape language studies taught us something valuable about the cognitive capacities of these magnificent animals, but very little about vocal communication.

An inescapable trend is increased use of digital processing methods. Just as sound spectrography changed the field over the past 25 years, the availability of data processing is changing it again. The capacity to handle large samples of vocal behavior, to improve signal to noise ratios, and to generate sets of parameters describing acoustic detail with great speed and reduced labor, is no longer limited to a few major institutions. Several of the reports in this volume represent technical advances which could not have been anticipated even 7 years ago. With this new sophistication, field data profit to the same extent as experimental approaches on captive animals.

In the organization of this book we have attempted to balance conceptual and methodological strengths. The intent has not been to present a technical manual, but to describe complex techniques where necessary in simple terms understandable to students and those readers who have an interest in the field but are competent in other disciplines.

A glossary is provided which explains the most important expressions and terms used throughout this volume. This is included with the intention to help newcomers in this area of research without discouraging him or her.

We take great satisfaction in the inclusion of contributions from field studies, which demonstrate that this approach is alive and well in the computer age. Sadly, the approach may be in better health than many primate species worthy of study, which are suffering grievously at this time from habitat destruction.

John Mitani and Peter Marler contrast the mechanisms of vocal communication in birds and primates. They conclude that manifold parallels exist, and that avian studies often have taken the lead in the discovery of principles of communication mechanisms. One example they provide is the recently found audience effect. On the other hand, primate communication features complex mechanisms which, such as the representational signaling, are less obvious in birds.

In a combination of data collection in the field and sophisticated analysis techniques in the laboratory, Jean-Pierre Gautier and Annie Gautier-Hion investigate as to whether, and to what extent, two sympatric West African forest guenons, *Cercopithecus nicticans* and *C. cephus*, differ as to acoustic features of their vocalizations. By means of parametrization of sound spectrograms and multivariate statistics they demonstrate that cohesion and contact calls could easily be discriminated, particularly by the rhythm of quavering. This rhythm is reported to be ontogenetically and contextually invariant and therefore appears to be a good candidate as a cue for individual recognition.

Wolf Dittus, reporting from his 2 decades of field experience with toque macaques, proposes a system of several cohesion calls guiding individuals in their selection of foraging partners and thereby contributing to foraging success. He also analyzes calls designating the status of being lost and/or seeking of contact. Special emphasis is given to the ontogenetic and ecological aspects of this communicative system of maintaining cohesion.

Charles H. Brown and Peter M. Waser review the available significance of natural habitat acoustics on structural features of primate vocalizations, as well as on perceptual capabilities. They give an overview of their classic work on the acoustic ecology of equatorial African primate habitats. Their chapter includes a report on their recent research on the way habitat acoustics influence the fidelity of calls broadcast in natural environments.

Communication is both a prerequisite and a manifestation of social organization. In this sense, the social aspect is implicitly inherent in all investigations studying communication. Several chapters in this volume, however, are dedicated to investigate the way vocal communication serves its ultimate function of maintaining social organization.

Manfred Maurus and his associates use a new analytic technique to discover the natural categories in primate vocal communication. They argue that appropriate functional categories of communicative processes can only be determined by "asking" the monkeys themselves what the essential components were that they respond to. In applying these techniques the authors show the great impact of amplitude modulation characteristics in squirrel monkey vocal communication.

The chapter provided by Dietmar Todt deals with communicative aspects of infant crying taken as a model of serial calling. Based on time series analysis of crying processes displayed by human babies and also by Barbary macaque infants, it is shown how the temporal dynamics of particular acoustic parameters affects responses of adult recipients. The results support the hypothesis that crying serves more than one social function.

Charles T. Snowdon places his focus on three complications necessary to be taken into account when considering communicative processes in the "real world": These are, firstly, the previous history of the relationship between communicants, secondly, the conversational nature of communication, and thirdly, the multidirectionality of communication in social environments. He supports the impact of these considerations with data from studies on ontogeny and interactive aspects of vocal exchanges in two callitrichid species, pygmy marmosets and cotton-top tamarins.

David Symmes and Maxeen Biben contribute a critical analysis of concepts dealing with reciprocal aspects in primate vocal exchanges. They suggest three criteria as the necessary evidence for conversation in subhuman primates: turn taking, directionality of change in acoustic structure, and bidirectionality of information transfer. The authors present quantitative data from studies in squirrel monkeys which provide clues for conversational behavior in this species.

Philipp Goedeking studies the covariance of several structural properties in cotton-top tamarin vocalizations with an increase in the rate of playful wrestling. He achieves this by applying computerized techniques of sound analysis and feature extraction. Covariation of behavioral and acoustical data are investigated by means of modified cross-correlation analyses. It is shown that, in this species, pitch instability, frequency range, and absolute pitch each covary with an increased rate of wrestle play.

Communication is mediated by a network of proximate mechanisms, all of which together form its biological substrate. Areas of research on the biological substrates of vocal behavior covered here include electrophysiology, neuroanatomy, and neuropharmacology.

Peter Müller-Preuss first reviews some common characteristics of the auditory pathway in mammals. He takes advantage of the evidence obtained from the work of Manfred Maurus and his associates on the significance of AM-envelope characteristics in squirrel monkey cackle calls. He substantiates Maurus' behavioral evidence to the extent that single cell recording experiments yielded examples of cells specifically responsive to such AM features.

A comparison of central control of speech and primate vocal production are the subject of Uwe Jürgens' chapter. He reviews evidence showing that monkey calls are genetically preprogrammed and are subject to very restricted voluntary control. These findings are supported by neuroanatomical studies investigating the brain structures involved in human speech control but dispensable for monkey call production and vice versa. Other brain structures such as the motoneurons innervating the phonatory muscles, the central gray of the midbrain, and the anterior limbic cortex are shown to be equally effective in human speech as well as monkey call production.

John Newman introduces ethopharmacology as a new means of access to understanding the proximate mechanisms mediating vocal behavior. He proposes the use of vocal behavior for studying drug action, since access to quantitative measurements of vocal structure potentially provides powerful data for interpreting effects of psychoactive drugs. Newman presents evidence of the significance of several drugs in mediating the vocal behavior of squirrel monkeys.

Two contributions in this volume are exceptional in that they develop conceptual frameworks of vocal communication. Both chapters, one by Klaus Scherer and Arvid Kappas, the other by Detlev Ploog, start from opposite ends of a phylogenetic perspective and emerge with essentially similar conclusions.

Klaus Scherer and Arvid Kappas present a theory on the effects of emotion on structural components in vocal utterances in man and subhuman primates. In essence, they propose that physiological states coupled with emotional variables push the vocal system into settings, and these, in turn, restrict structural characteristics of possible vocal output. Predictions of changes in vocal output, from a proposed model of serial evaluation of stimuli in affective situations, allow a comparative match with data from their own work on human speech and published data on primate behavior.

Detlev Ploog contributes an in-depth review of clinical evidence on the mechanisms of central nervous control of vocal production in man and monkeys. Systematic studies of pathological deficits in the ability to speak, along with results obtained by brain lesion and brain stimulation experiments in monkeys, allow a well-grounded comparative perspective. Ploog concludes that, despite significant fundamental differences, several prerequisites of human speech are present in the vocal behavior, and its central control, in subhuman primates.

Berlin and Bethesda, Summer 1988

D. TODT
P. GOEDEKING
D. SYMMES

Contents

Biological Substrates

Conceptual Frameworks

Contributors

You will find the addresses at the beginning of the respective contribution

Comparative and Field Studies

Vocal Communication in Primates and Birds: Parallels and Contrasts

PETER MARLER and JOHN MITANI[1]

During the past few years primatologists and ornithologists have joined forces to address common problems in the study of vocal communication. In our own research we have made a special point of employing broadly comparative approaches to the investigation of functional and developmental aspects of primate and avian vocal signaling. In this chapter we outline the rationale we have adopted, report some of the results of our research, and indicate where ethologists investigating avian and primate vocal systems may benefit from mutual awareness of each other's research.

1 Social Systems and Vocal Communication

Insofar as communication processes mediate social interactions, variations in social systems will exert a strong influence on patterns of vocal signaling. Primates and birds exhibit a variety of social systems, ranging from monogamous, territorial pairs to multimale – female groups and colonies (Wittenberger 1979). Among birds, the widespread phenomenon of territoriality places a premium on long-range signaling. As a result, bird song is a major feature of life in many avian species and continues to be a focus of research for ornithologists (reviews in Kroodsma and Miller 1982). In contrast, large social groupings predominate among nonhuman primate species, and here the attendant communication problems faced by animals are quite different than those of monogamous, territorial birds. Many primatologists have consequently investigated short-range vocal signals which regulate subtle, within-group social interactions (reviews in Snowdon et al. 1982).

As research efforts in avian and primate communication have progressed on different fronts, we have found that advances in one field serve as the basis for asking questions in the other. Specifically, research on male bird song has provided a model for functional investigations of loud vocalizations given by primates, while recent progress in the study of the semantic communication systems of nonhuman primates has served as an impetus to search for analogous signaling systems in birds.

1 The Rockefeller University Field Research Center, Tyrrel Road, Millbrook, New York 12545, USA

Todt/Goedeking/Symmes (Eds.)
Primate Vocal Communication

2 The Function of Loud Vocalizations

In most oscine bird species there is a correlation between the onset of male singing and the emergence of territoriality, pair formation, and mating. This correlation has led to the hypothesis that song plays a role in territorial defense and mate attraction. Field playback studies using tape recorded song have recently provided the first direct evidence supporting these hypotheses (Krebs 1977; Yasukawa 1981; Eriksson and Wallin 1986).

The roles of bird song in spacing and mating behavior are now well established, but although a spacing function for the loud calls of male primates has long been assumed (Marler 1968), our understanding of the functions of these vocalizations has lagged behind. Recent progress has been made, however, as primatologists have adopted the field playback methods refined by ornithologists. Early attempts to employ the sound playback technique with nonhuman primates failed, including some efforts of our own research group, perhaps because animals were permitted to identify the playback source and to habituate to the test signal (Green 1975). By controlling experimental conditions carefully, Waser (1975) was the first to use playbacks successfully in the field to determine the function of a nonhuman primate vocalization.

Subsequent to Waser's pioneering research, several investigators have utilized the playback procedure to investigate the function of a variety of primate vocal signals (Seyfarth in Smuts et al. 1987). Following the lead provided by ornithologists, most research has focused on long-range signals, hypothesized to play a role in mediating spacing and mating behavior. Fieldwork on the vocal behavior of gibbons and orangutans illustrates the insights to be derived from these studies.

Gibbons are small arboreal apes that inhabit the rain forests of Southeast Asia. Social groups, consisting of permanently mated adult pairs and their offspring, defend site-specific territories. In addition to territoriality and monogamy, gibbons exhibit remarkably birdlike singing abilities. Mated and unmated male gibbons engage in solitary singing performances, while mated pairs accompany each other to produce elaborate duets. Several investigators, beginning with Carpenter (1940), have hypothesized that singing by gibbons performs a spacing function similar to bird song. Direct evidence supporting this hypothesis depends on showing that song influences the movements of conspecific animals in predictable ways (Marler 1968).

To investigate this possibility, playback experiments were conducted using recorded male songs and duets of grey gibbons (Mitani 1985a,b). Results of these experiments showed that the gibbons do indeed respond to song playbacks, that habituation can be avoided by taking appropriate precautions and that the animals modulate their responses as a function of the location of the playback speaker. Male songs and duets elicited approaches when played back from the center of a test group's territory. In contrast, playbacks emanating from the group's territorial boundary evoked singing responses only; songs from an adjacent neighbor's territory rarely resulted in approaches or song.

While these data are consistent with the hypothesis that gibbon song serves a spacing function, others have pressed the bird song analogy farther by proposing that unmated male gibbons use song to attract mates (e.g., Tenaza 1976). This hypothesis overlooks an important difference in the biology of the species under comparison, however. Most temperate region male birds breed at regular seasonal intervals and declare their unmated status at these times by singing. Here the unmated status of singing males is indicated simply by a temporal cue, the start of the breeding season, and the marked reduction in singing rates that sometimes occurs after mating. In contrast, male gibbons remain permanently mated, with pair bonds that endure for several years. The mate attraction hypothesis rests on the assumptions that the songs of unmated and mated males are distinct and that females can differentiate them.

In a preliminary analysis, no consistent differences in the qualitative acoustic morphology of the songs of mated and unmated male gibbons have been detected. In quantitative terms, however, the songs of unmated males appear to be longer and sung at a faster rate than those of mated males. In recent playback studies, tests were conducted to determine whether gibbons use these variations in temporal features to differentiate mated and unmated male song (Mitani, unpublished data). Results showed that the gibbons did not respond differently to the songs of mated and unmated males. No significant differences existed in approach or vocal responses. In addition, quantitative variation in approach response was not revealed; the latency to approach mated and unmated male songs did not differ. More importantly, subadult females in five groups tested did not react at all to the playback of male song. These results suggest that gibbons do not differentiate unmated from mated male song and that unmated males would have difficulty in attracting prospective mates through the use of song alone.

In contrast to gibbons, orangutans are primarily solitary; the only enduring social groupings consist of females and their offspring. The strikingly different social systems of orangutans and gibbons have provided an ideal opportunity to explore the possible spacing and mating functions of long-range primate vocal signals in comparative perspective. Adult male orangutans emit loud, sex-specific vocalizations termed "long calls". Previous field-workers have hypothesized that these calls mediate interindividual spacing between males and are used by females to locate high-ranking mates over long distances (Rodman and Mitani in Smuts et al. 1987). To test these two hypotheses, field playback experiments were conducted using tape recorded calls (Mitani 1985c). Experiments showed that long calls regulate spacing between males through an approach-avoidance system based on dominance relationships; the highest ranking adult male approached calls, while low ranking males avoided calls. Playbacks also revealed no tendency for sexually active females to move toward long calls, and thus did not support the mate attraction hypothesis.

The experimental results presented here suggest that in some cases the loud calls of male primates serve important spacing functions, but unlike male bird song, do not play a role in attracting mates. This difference illustrates the lack of complete functional equivalence between avian and primate vocal systems. Future studies of primate vocal behavior should seriously entertain alternative interpretations of signal function (e.g., Sekulic 1982).

3 Species Recognition

Species differences in vocal behavior have been noted frequently by ornithologists, dating back to Gilbert White's use of song in the 18th century to distinguish species of

European warblers. Recently, primatologists have employed the loud vocalizations given by many adult male forest primates to differentiate morphologically similar populations of old and new world monkeys (Struhsaker 1970; Oates and Trocco 1983; Snowdon et al. 1986).

Playback experiments provide direct evidence for specific recognition of songs in birds (Becker in Kroodsma and Miller 1982), but comparisons of responsiveness to conspecific and closely related heterospecific signals have rarely been performed with nonhuman primates in the field. To investigate species recognition of vocalizations in a nonhuman primate, we recently tested whether gibbons discriminate between the songs of different gibbon species. For these experiments male songs from two morphologically similar species, the agile gibbon and the grey gibbon, were used. Songs were recorded from three distinct geographic populations, West Kalimantan and South Sumatran agile gibbons and East Kalimantan grey gibbons, and played back to a population of agile gibbons in West Kalimantan. Only slight differences were found in the acoustic structure of agile gibbon songs from West Kalimantan and South Sumatra, but both differ strongly from the songs of male grey gibbons in East Kalimantan. The experiments showed that the gibbons responded to local song and conspecific South Sumatran song by approaching silently. In contrast, grey gibbon male song did not reliably elicit approach responses, but instead evoked calling responses by females. Studies of gibbons (see also Raemaekers and Raemaekers 1985) provide the only field evidence for specific recognition of vocalizations in nonhuman primates and point to the need for more work in this area.

4 Individual Recognition

Individual differences in the vocalizations of birds, readily apparent to even the casual observer, have received considerable study by ornithologists. Playback experiments routinely reveal that birds are able to recognize the vocalizations of individual conspecifics (Falls in Kroodsma and Miller 1982).

As is the case in bird vocalizations, striking variations often exist among the calls of individual primates. There are strong theoretical reasons for believing that this variability is biologically meaningful. Most primates live within complex social groups and interact daily with several individuals, varying in such factors as kin relationship and rank. Since vocalizations often mediate these interactions, there must be a high premium on the ability to identify the calls of different group members.

To investigate whether monkeys and apes recognize companions individually by voice alone, primatologists, again following the lead provided by ornithologists, have employed the experimental playback technique. Among primates mother-infant bonds are critical and enduring; playback studies have shown that females and infants recognize each others' calls (Kaplan et al. 1978; Cheney and Seyfarth 1982a; Masataka 1985; Symmes and Biben 1986). Similarly, individual recognition of vocalizations used to mediate within-group relations in neotropical callitrichids has been demonstrated experimentally by Snowdon and colleagues (Snowdon in Snowdon et al. 1982).

Recognition of vocal signals is not limited to those used for communication within the confines of the social group. Birds have been commonly shown to discriminate between the territorial songs of familiar and unfamiliar males (Falls in Kroodsma and Miller 1982), but individual recognition of long-distance vocalizations by primates in the field has been more difficult to document. Although the evidence on individual discrimination of song duets by gibbons is negative thus far (Mitani 1985d), individual recognition of loud vocalizations has been demonstrated in mangabeys (Waser 1977) and in vervets (Cheney and Seyfarth 1982a).

5 The Acoustic Structure of Repertoires

In the past considerable significance has been attached to an apparent contrast in the nature of vocal signal variation in birds and primates. Birds were thought to have repertoires consisting mainly of discrete signals. In higher primates highly graded signals were thought to predominate, such that discrete categories are often difficult to discern (e.g., Marler 1975). It has become increasingly clear that this is an oversimplification.

The classic case of graded signal structure is the rhesus macaque, whose vocalizations were first described by sound spectrography in a pioneering study by Rowell (1962). It was concluded that much of the vocal repertoire is organized as a graded system, and that the visual signals that accompany them are the primary method of communication. One major component in this vocal system consisted of scream calls.

This problem has recently been reanalyzed (Gouzoules et al. 1984), focusing on rhesus calls given during agonistic encounters. These calls were found to be organized into relatively discrete categories, comprising five major classes. Of 561 scream bouts analyzed, 90.4% (507) contained sounds that clearly belonged to one of the five broad classes. All but 9.62% of the screaming bouts could be unambiguously classified in this way. Interestingly, as noted by Green (1975) in the Japanese macaque, the calls also tended to be given in bouts of a single type. Only 16% of the bouts incorporated more than one scream class. It was still less common to find an intergraded series between two scream classes within a given bout. Thus even this classic example of signal intergrading proves on closer investigation to have a discrete organization.

This is not of course to say that rhesus calls are completely stereotyped. A great deal of grading between variants within a given call class was found, but this was much more frequent than grading between classes. Prominent and consistent individual differences occurred within each class, sufficient to have communicative significance, but these did not hinder placement of calls into the same limited set of relatively discrete categories. There were striking variations in the frequency of usage of some calls according to sex and age but all five classes nevertheless occur in the vocal repertoire of a single individual.

It may well be that further quantitative analysis of other primate vocal systems presumed to be highly intergraded will reveal a discrete organization. Conversely, the generalization that most avian vocal repertoires are discretely organized may be invalid, and will repay further investigation. Graded vocal signals may in fact prove to be as widespread in birds as in primates, provided that they live in groups with complex patterns of social organization (e.g., Maier 1982, Nelson 1985). The relationship between highly variable signal morphology, use at close range, cooperation between auditory and visual signals, and occurrence in the context of complex social interrelationships may hold in both birds and nonhuman primates (Marler 1975).

6 Vocalizations as Representational Signals

The most striking recent advance in our understanding of the function of vocal signals in nonhuman primates concerns their use in a representational manner. At the time of discovery, this had not been described previously in avian communication. There are at least two well-documented cases of primates apparently using vocal signals to represent objects in the environment. First, there is evidence that the toque macaque has a class of calls representing food (Dittus 1984). Second, the vervet monkey has a large repertoire of alarm calls, some of which appear to represent different kinds of predators (Struhsaker 1967). More recently referential signaling about social factors, such as kinship and rank, has been shown. Vervet monkeys have a variety of grunts occurring in different social situations. Field observations and playbacks revealed that the type of grunt given when approaching a dominant companion elicits different responses than the grunt type given when approaching a subordinate, and different again from grunts given when a companion initiates movement of the social group away from cover, or sights an adjacent group (Cheney and Seyfarth 1982b).

Another case of representational signaling about social relationships going beyond anything yet described in birds concerns the scream calls of rhesus macaques used to recruit aid from nearby relatives in the course of a fight (Gouzoules et al. 1984). Playbacks of screams recorded from juveniles in a variety of agonistic situations to groups of adults that included the callers' mothers yielded convincing evidence that each of the five call types used represents a different combination of social and physical factors. These appeared to include genealogical relationships between the caller and its opponent, dominance relationships, and the risk of harm to the caller (Gouzoules et al. 1984).

The revelation that some primates use vocalizations in a representational fashion led to reinvestigation of this issue in birds, using chickens as subjects (Marler et al. 1986a). Many gallinaceous birds produce calls when they find food, particularly in heterosexual courtship feeding, and when providing food for the young. Laboratory experiments showed convincingly that cockerels have a call that is associated with discovering food and that call rate and number vary according to the preference ranking of the particular food item. If a hen is present, she will approach a food-calling male and accept the food he proffers, and is more likely to do so than when he is silent. If he calls he is more likely than not to refrain from eating himself, and to give the hen prior access, even if it is a highly preferred food. Most interestingly, the probability of her approach is greater when the male calls to a high preference food, as compared with a low preference item (Marler et al. 1986a). From this research it now appears that both avian and primate signals can operate in a representational mode (see also Elgar 1986).

Evidence of representational signaling has also been uncovered in the alarm calls of chickens. Ground-predator alarm calls of cockerels are given with a high degree of specificity to objects moving on the substrate. They are acoustically quite distinct from aerial alarm calls elicited by objects moving above in free space (Gyger et al. 1968). Further correlations between call structure and the precise nature of the danger stimulus have been found, less closely linked to the specific identity of the aerial stimulus than to its size and rate of movement. This lack of specificity is construed, not as evidence of a more "primitive" or nonrepresentational system, but as the result of a compromise.

The costs of sometimes alarming to nonpredators are probably counterbalanced by the benefits of an expanded field of vision, but this limits binocularity, impedes distance judgements, and reduces the accuracy of object identification. Only further work will show whether overflying birds of different species can be discriminated when they subtend the same visual angle, or whether they are discriminable but grouped together for purposes of alarm call production. It is important to bear in mind that objects can be classified in a multitude of different ways, not necessarily coincident with those that appear most natural to us, while still providing the basis for a representational system of communication.

One intriguing fact, well known to students of domestic chicken behavior, is that some so-called food signals are produced in response to inedible objects, especially during the "tidbitting" phase of courtship (e.g., Kruijt 1964). Further investigations of the behavior of chickens in large outdoor pens (Gyger and Marler 1988) support many of the laboratory findings. They also yielded quantitative data on the frequency of food calling without food, which made up a total of 45% of all food calling. No difference could be detected in the structure of male food calling with and without food. Analysis of female responses revealed that the probability of female approaches was correlated with male calling rates in exactly the same way as when food was present. The major contrast was that calling without food occurred at significantly greater distances between cockerel and hen than calling with food, suggesting that this may be a case of deception. In a number of nonfood-calling cases, the approaching female scratched or pecked at the ground, as though she expected to find food there.

If animals are indeed capable of signaling in a representational fashion, the possibility of deceptive use is an important corollary, likely to be manifest either by proferring information that is false or by withholding information that is true (Smith 1985). Munn (1986) has described another case of apparent deception in birds, based on the use of alarm calls in the absence of predators. For the moment, research on the deceptive use of natural vocal signals appears to have progressed further with birds than with primate subjects, although there are indications of deceptive use of alarm calls by vervet monkeys under field conditions (Cheney et al. 1986).

7 The Audience Effect

The importance of alarm calls in affecting the survival of close kin (Klump and Shalter 1984; Sherman 1985) should prepare us for the possibility that, in dangerous situations, signals might be emitted in the presence of appropriate receivers, but withheld in their absence. This would presume a degree of control over production of alarm calls. In their studies of responses of vervet monkeys to stimuli signifying danger, Cheney and Seyfarth (1985) found that captive females gave more alarm calls in the presence of their own infant than with an unrelated infant. Effects of the presence of an appropriate audience for signal emission have been explored in most detail in birds, revealing some remarkably subtle effects.

Food-calling cockerels are acutely sensitive to the presence and nature of the audience (Marler et al. 1986b). They produce significantly more food calls in the presence of a strange female than with a familiar female, and still less with no audience at all.

A baseline level of food calling persists without an audience, and even this was inhibited if another male was present.

Subsequent work has shown that alarm calling by cockerels is also affected strongly by an audience (Gyger et al. 1986). Working in the laboratory and using aerial models of raptors as stimuli, males called more in the presence of the mate or another female than when they were alone. In contrast to food calling, however, presenting another male as an audience facilitated alarm calling rather than inhibiting it. Members of other species had no effect.

Audience gender is important for male food calling but not for aerial alarm calling. One explanation for the "audience effect" is that there is mutual facilitation of alarm behavior because both are responding to the same danger stimulus. The effect persists, however, if the audience is prevented from seeing the danger stimulus presented overhead. The contrast between audience effects on food and alarm calling suggests that something other than arousal by a companion's presence is involved. The nature of the effect of an audience is not uniform across different communicative situations, but varies from system to system in specific ways.

The "audience effect" may provide an empirical means by which we can begin to explore the nature and limits of the elusive and yet fascinating subject of intentionality in animal communication (Dennett 1983). Given the complexities of primate sociality, this approach may be even more fruitful when applied to monkeys and apes. If we can demonstrate that both birds and nonhuman primates are able, by design, to manipulate the behavior of others in a dynamic fashion by subtle control of signal production, we shall have to entertain seriously the proposition that intentionality is not uniquely human.

8 Vocal Learning

The ontogeny of bird song is an active area of current ornithological research. Laboratory and field evidence reveals that learning plays a major role in the development of the production of male bird song (Kroodsma in Kroodsma and Miller 1982). In the laboratory, young oscines raised in acoustic isolation from conspecific sounds typically develop abnormal song; in the field, neighboring males frequently share local song dialects.

Attempts to show vocal learning in primates using the methods employed in bird song research have produced mixed results (see Snowdon, this Vol.). The inconclusive results of these studies have encouraged other approaches to the study of primate vocal ontogeny. For example, hybridization studies in which first generation crosses and backcross offspring are shown to possess vocalizations unlike either of their parents, and with no obvious resemblance to sounds of other species within earshot, are frequently cited as evidence that the development of primate vocalizations is largely under genetic control (e.g., Brockelman and Schilling 1984).

Although the results of this research are interesting, we nevertheless believe that these data should be viewed cautiously when attempting to evaluate the role of learning in the ontogeny of natural primate vocal behavior. If innate genetic predispositions are involved in any learning that does take place, as is the case in some birds (e.g., Marler and Peters 1977, 1988), their contributions to vocal development in hybrids are likely to be complex and difficult to appraise without further experimentation. Thus a hybrid might fail to learn parental vocalizations, not because it is incapable of learning, but because its innate predispositions are not well matched to the sounds to which it is exposed.

The lack of robust evidence implicating learning as an important factor in the development of primate vocal production is puzzling. One explanation seldom entertained is that acoustic changes in vocalizations resulting from learning are so subtle that they are liable to be overlooked. This may be especially true of the broad-band, noisy signals characteristic of many primates, spectrograms of these sounds are often difficult to interpret by eye. The application of digital signal-processing techniques (e.g., Goedeking 1983; Clark et al. 1988), sensitive to variations in spectral structure, may aid in the description of such signals and provide new insights into the development of nonhuman primate vocal production.

While avian studies have concentrated on the ontogeny of vocal production, the developmental processes by which birds learn to sing in appropriate circumstances and to respond properly to song have received comparatively little study. More progress has been made in our understanding of the development of primate vocal usage and responsiveness to calls. Field research on the development of vervet monkey vocalizations has shown that young animals gradually sharpen their ability to produce two sets of calls, alarms and grunts, in their proper contexts, as they mature (Seyfarth and Cheney 1986). Moreover, the responsiveness of infant vervets to alarm calls follows a similar gradual developmental time course; infants do not always respond properly to alarm calls, but are more likely to do so if they first receive cues from older group members (Seyfarth and Cheney 1986). Recent laboratory research on Goeldi's monkeys and Japanese macaques complements these findings by showing that the responsiveness of infants to conspecific alarm calls and to the calls of their mothers develops over time (Masataka 1983, 1985).

In summary, controversy about the contributions of learning to primate vocal development will only be resolved by rearing subjects under controlled conditions. It would be especially illuminating to include hybrids and cross-fostered infants in experiments of this kind. Recent applications of digital signal-processing techniques and the fine-grain descriptions of primate and avian vocalizations they generate promise to lead to a better understanding of both primate and avian vocal development. The ontogeny of both vocal usage and vocal responsiveness, in primates and birds, merits closter attention than it has received in the past.

9 Conclusions

Only 10 years ago, the proposal that radically new insights into primate vocal communication might emerge from experimental playback of vocal signals under field conditions would have been greeted with skepticism. The fact that bird song playbacks had already provided novel viewpoints about the communication systems of birds would have been no guarantee that more intelligent animals, such as monkeys and apes, would be gullible enough even to respond consistently to vocal signal playbacks. It is now clear that when appropriate precautions are taken to minimize habituation and to simulate natural conditions as closely as possible, the vocal playback technique is a sensitive method for asking penetrating questions about the communicative behavior of non-human primates, questions that would be difficult to pose by other experimental approaches. Thus vocal signal playback has been used to demonstrate that, as in birds, loud calls of higher primates play a role in inter-group and -individual spacing. Proof of a sexual function has been more elusive, although this is an issue that merits further investigation. Recognition of both species and individuals is mediated by vocal signals in many primates, again echoing what had been demonstrated previously in birds.

More novel and unexpected was the revelation that some primate vocal signals function in a representational fashion, serving to communicate information not only about objects in their environment but also about more abstract phenomena such as social relationships. Nothing equivalent has been shown or even postulated in birds, to our knowledge. Avian studies have taken the lead, however, with the discovery of the so-called audience effect. Some birds are able to modulate production of food and alarm calls according to whether or not an appropriate signal receiver is present, hinting at a new way of approaching the challenging issue of intentionality in animal behavior. Primates, with their intricate social networks, are obvious subjects for further experimentation on this issue.

There is a particularly urgent need for deeper analyses of vocal communication in the anthropoid apes, especially the chimpanzee and the bonobo. Given the complexity of their social systems and the remarkable cognitive abilities revealed by laboratory experimentation, it seems evident that we have hardly begun to appreciate the subtle communicative interactions between these animals that must be taking place in nature. Continuing theoretical, methodological, and empirical interchanges between primatologists and ornithologists hold promise for future studies into the relationships between vocal communication, cognition, and social relationships in these two species.

Acknowledgments. Research reported here was supported by a U.S.P.H.S. postdoctoral fellowship #2 F32 NSO7670, grants from the L.S.B. Leakey Foundation and the National Geographic Society to J.M., and a grant from the National Science Foundation #BNS 8416451 to P.M. Fieldwork was sponsored by the Indonesian Institute of Sciences, the Indonesian National Biological Institute, and the Indonesian Directorate of Forest Protection and Nature Conservation. We thank Drs. Douglas Nelson, Stephen Nowicki, and Marc Hauser for helpful discussion and comments on the manuscript and Esther Arruza for patiently typing it.

References

Brockelman W, Schilling D (1984) Inheritance of stereotyped gibbon calls. Nature (London) 312: 634–636

Carpenter CR (1940) A field study in Siam of the behavior and social relations of the gibbon *(Hylobates lar)*. Comp Psych Monogr 16:1–212

Cheney D, Seyfarth R (1982a) Recognition of individuals within and between groups of free-ranging vervet monkeys. Am Zool 22:519–529

Cheney D, Seyfarth R (1982b) How vervet monkeys perceive their grunts. Anim Behav 30:739–751

Cheney D, Seyfarth R (1985) Vervet monkey alarm calls: manipulation through shared information? Behaviour 94:150–166

Cheney D, Seyfarth R (1986) Vocal development in vervet monkeys. Anim Behav 34:1640–1658

Cheney D, Seyfarth R, Smuts B (1986) Social relationships and social cognition in nonhuman primates. Science 234:1361–1366
Clark C, Marler P, Beeman K (1988) Quantitative analysis of animal vocal phonology: an application to swamp sparrow song. Ethology 76:101–115
Dennett D (1983) Intentional systems in cognitive ethology: the "Panglossian paradigm" defended. Behav Brain Sci 6:343–390
Dittus W (1984) Toque macaque food calls: semantic communication concerning food distribution in the environment. Anim Behav 32:470–477
Elgar M (1986) House sparrows establish foraging flocks by giving chirrup calls if the resources are divisible. Anim Behav 34:169–174
Eriksson D, Wallin L (1986) Male bird song attracts females – a field experiment. Behav Ecol Sociobiol 19:297–299
Goedeking P (1983) A minicomputer-aided method for the detection of features from vocalizations of the cotton-top tamarin. Z Tierpsychol 62:321–328
Gouzoules S, Gouzoules H, Marler P (1984) Rhesus monkey *(Macaca mulatta)* screams: representational signalling in the recruitment of agonistic aid. Anim Behav 32:182–193
Green S (1975) Variation of vocal pattern with social situation in the Japanese monkey *(Macaca fuscata)*: a field study. In: Rosenblum L (ed) Primate behavior. Academic Press, London New York, pp 1–102
Gyger M, Marler P (1988) Food calling in the domestic fowl *(Gallus gallus)*: the role of external referents and deception. Anim Behav (in press)
Gyger M, Karakashian S, Marler P (1986) Avian alarm calling: is there an audience effect? Anim Behav 34:1570–1572
Gyger M, Marler P, Pickert R (1988) Semantics of an avian alarm call system: the male domestic fowl, *Gallus domesticus*. Behaviour 36:358–365
Kaplan J, Winship-Ball A, Sim L (1978) Maternal discrimination of infant vocalizations in squirrel monkeys. Primates 19:187–193
Klump G, Shalter M (1984) Acoustic behavior of birds and mammals in the predator context. Z Tierpsychol 66:189–226
Krebs J (1977) Song and territory in the great tit. In: Stonehouse B, Perrins C (eds) Evolutionary ecology. University Park Press, Baltimore, pp 47–62
Kroodsma D, Miller E (1982) Acoustic communication in birds, vols 1, 2. Academic Press, London New York
Kruijt JP (1964) Ontogeny of social behaviour in Burmese red jungle fowl (*Gallus gallus spadiceus* Bonaterre). Behaviour Suppl 12:1–201
Maier V (1982) Acoustic communication in the Guinea fowl *(Numida meleagris)*: structure and use of vocalizations and the principles of message coding. Z Tierpsychol 59:29–83
Marler P (1968) Aggregation and dispersal: two functions in primate communication. In: Jay P (ed) Primates. Holt, Rhinehart, Winston, New York, pp 420–438
Marler P (1975) On the origin of speech from animal sounds. In: Kavanaugh J, Cutting J (eds) The role of speech in language. MIT, Cambridge, pp 11–37
Marler P, Peters S (1977) Selective vocal learning in a sparrow. Science 198:519–521
Marler P, Peters S (1988) The role of song phonology and syntax in vocal learning preferences in the song sparrow, *Melospiza melodia*. Ethology 77:125–149
Marler P, Dufty A, Pickert R (1986a) Vocal communication in the domestic chicken: I. Does a sender communicate information about the quality of a food referent to a receiver? Anim Behav 34:188–193
Marler P, Dufty A, Pickert R (1986b) Vocal communication in the domestic chicken: II. Is a sender sensitive to the presence and nature of a receiver? Anim Behav 34:194–198
Masataka N (1983) Categorial responses to natural and synthesized alarm calls in Goeldi's monkeys *(Callimico goeldi)*. Primates 24:40–51
Masataka N (1985) Development of vocal recognition of mothers in infant Japanese macaques. Dev Psychobiol 7:351–358
Mitani J (1985a) Gibbon song duets and intergroup spacing. Behaviour 92:59–96
Mitani J (1985b) Location-specific responses of gibbons *(Hylobates muelleri)* to male songs. Z Tierpsychol 70:219–224

Mitani J (1985c) Sexual selection and adult male orangutan long calls. Anim Behav 33:272–283

Mitani J (1985d) Responses of gibbons *(Hylobates muelleri)* to self, neighbor, and stranger song duets. Int J Primatol 6:193–200

Munn C (1986) Birds that 'cry wolf.' Nature (London) 319:143–145

Nelson D (1985) The syntactic and semantic organization of pigeon guillemont *(Cepphus columba)* vocal behavior. Z Tierpsychol 67:97–130

Oates J, Trocco T (1983) Taxonomy and phylogeny of black-and-white colobus monkeys. Inferences from an analysis of loud call variation. Folia Primatol 40:83–113

Raemaekers J, Raemaekers P (1985) Field playback of loud calls to gibbons *(Hylobates lar)*: territorial, sex-specific and species-specific responses. Anim Behav 33:481–493

Rowell T (1962) Agonistic noises of the rhesus monkey *(Macaca mulatta)*. Symp Zool Soc London 8:91–96

Sekulic R (1982) The function of howling in red howler monkeys *(Alouatta seniculus)*. Behaviour 81:38–54

Seyfarth R, Cheney D (1986) Vocal development in vervet monkeys. Anim Behav 34:1640–1665

Seyfarth R, Cheney D, Marler P (1980) Vervet monkey alarm calls: semantic communication in a free-ranging primate. Anim Behav 28:1070–1094

Sherman P (1985) Alarm calls of Belding's ground squirrels to aerial predators: nepotism or self-preservation? Behav Ecol Sociobiol 17:313–323

Smith WJ (1985) An "informational" perspective on manipulation. In: Mitchell R, Thompson N (eds) Deception. State Univ New York Press, Albany, pp 71–86

Smuts B, Cheney D, Seyfarth R, Wrangham R, Struhsaker T (1987) Primate societies. Chicago Univ Press, Chicago

Snowdon C, Brown C, Petersen M (1982) Primate communication. Cambridge Univ Press, Cambridge

Snowdon C, Hodun A, Rosenberger A, Coimbra-Filho A (1986) Long-call structure and its relation to taxonomy in lion tamarins. Am J Primatol 11:253–262

Struhsaker T (1967) Auditory communication among vervet monkeys *(Cercopithecus aethiops)*. In: Altmann S (ed) Social communication among primates. Chicago Univ Press, Chicago, pp 281–324

Struhsaker T (1970) Phylogenetic implications of some vocalizations of Cercopithecus monkeys. In: Napier J, Napier P (eds) Old world monkeys. Academic Press, London New York, pp 365–444

Symmes D, Biben M (1986) Maternal recognition of individual infant squirrel monkeys from isolation call playbacks. Am J Primatol 9:39–46

Tenaza R (1976) Songs, choruses and countersinging of Kloss' gibbons *(Hylobates klossi)* in Siberut Island, Indonesia. Z Tierpsychol 40:37–52

Waser P (1975) Experimental playbacks show vocal mediation of intergroup avoidance in a forest monkey. Nature (London) 255:56–58

Waser P (1977) Individual recognition, intragroup cohesion and intergroup spacing: evidence from sound playback to forest monkeys. Behaviour 60:28–74

Wittenberger J (1979) The evolution of mating systems in birds and mammals. In: Marler P, Vandenbergh J (eds) Handbook of behavioral neurobiology, vol 3. Plenum, New York London, pp 271–349

Yasukawa K (1981) Song and territory defense in the red-winged blackbird. Auk 98:185–187

Vocal Quavering: A Basis for Recognition in Forest Guenons

JEAN-PIERRE GAUTIER and ANNIE GAUTIER-HION [1]

Cercopithecus nictitans and *C. cephus* (Cercopithecinae), two guenons of West African tropical forests, are charăcterized by a tendency to live together in mixed troops (Gautier and Gautier-Hion 1969). In such sympatric troops, individuals of the two species vocalize in common vocal bouts regardless of the context of emission: alarm situations, troop progression, intertroop spacing, etc. The acoustic structure of most of their calls is so similar that species are very often difficult to identify by an observer relying on calls only. In fact, loud calls given by adult males are the only sounds which allow unambiguous species recognition (Gautier 1975).

This similarity in vocal repertoires agrees with findings from karyological and blood protein studies (Dutrillaux et al. 1982; Ruvolo 1988) and reflects close phylogenetic affinities between the two species (Gautier 1988). However, within stable polyspecific troops such as those found in Gabon (Gautier-Hion and Gautier 1974), troop members of every species have to maintain intraspecific cohesion and one might expect species recognition in the case of cohesion calls given without the support of visual cues, whereas such recognition might or might not be found for alarm calls.

A detailed analysis of equivalent calls given by *C. nictitans* and *C. cephus* was thus conducted to examine the extent to which their acoustic parameters differ. We also wished to discover which parameters may play important roles in species and individual recognition.

1 Material and Methods

The structural and functional analyses of the vocal repertoires of *C. nictitans* and *C. cephus* have been published elsewhere (Gautier 1975, 1978). For the present study, large samples of calls belonging to two functional categories were recorded either from wild Gabonese monkeys (the case for the overall study of acoustic features) or from selected animals whose age and sex were precisely known. The latter, maintained in captivity in our laboratory, came from the same population in Gabon.

Recordings were made with Nagra IV L and 4-2 tape recorders at a tape speed of 7.5 in. s^{-1} (bandwidth 30 Hz to 15 kHz ± 2 dB) using the highly directional condenser microphone Sennheiser MKH 815.

In the field, cohesion/contact calls were recorded (1) early in the morning and late afternoon near the sleeping sites, (2) during daily locomotor activities, and (3) when

1 Université de Rennes I, Station Biologique de Paimpont, 35380 Plélan le Grand, France

Todt/Goedeking/Symmes (Eds.)
Primate Vocal Communication

monkeys were collecting fruits in trees. The distance of such recordings ranged from 10 to 30 m. In captivity, recordings were made at distances less than 10 m.

Spectrographic analyses were made with the Kay sonagraph (7029 A & 6061 B) after relevant filtering of original tapes (especially for field recordings). The full scale (80 Hz/8 kHz) and the wide-band filter (300 Hz) were used for the majority of spectra. The narrow-band filter (45 Hz) and a magnified scale (40 Hz/4 kHz) were used to detect the pitch of the low-pitched cohesion calls. Parameters which were measured are those commonly used in acoustic studies such as: duration (ms); upper and lower frequencies (kHz); "dominant frequency" (kHz) measured at the midpoint of what appears to be the blackest part of the call's spectrogram (Marler 1973); "amplitude frequency" (kHz); "slope" (kHz s^{-1}) and pitch which characterize the fundamental tone; and the "rhythm of the quavering" (Hz). For high- and medium-pitched trills the latter is the ratio of the number of complete oscillations of the fundamental frequency on their total duration. For the discontinuous low-pitched cohesion calls, it is the inverse value of the mean period of the succession of subunits of a call (see Gautier 1978, p. 126).

The two vocal types envisaged include calls given in the context of alarm and calls playing a role in intratroop cohesion and more refined interindividual relationships. The structure of the sounds given by the two species are almost completely analogous, so each vocal type is characterized by the same set of acoustic parameters. The latter can be compared one by one or together. The t-test and the approximated t-test (Sokal and Rohlf 1969) were used in the first case, and a multifactorial correspondence analysis and two discriminant analyses (Benzécri 1973; Fénelon 1981; Gouet et al. 1985; Bachacou et al. 1981) in the second. These latter procedures permit us to estimate for each vocal type the level of similarity or dissimilarity between the two species and to state which parameters contribute most to the species discrimination.

The correspondence analysis (here STAT-ITCF, Gouet et al. 1985) seeks to extract, from a cloud of more or less dependent points in an n-dimensional space, the independent orthogonal axes with maximum inertia that account for the most variance in a smaller space than that occupied by the initial points. The projection of the cloud of points onto the factorial plane, which possesses the greatest variance (inertia), allows us to visualize the localization of data for each species, to estimate their level of distinctiveness, and to disclose the involved parameters. The closer to one axis the parameters are located, the more they contribute to its orientation.

Two discriminant analyses have been used which classify data for each vocal type given by the two species, according to their acoustic parameters. They provide both information about the parameters on which the discrimination most probably relies and estimate their relative weight. One of these analyses (AMANCE, Bachacou et al. 1981) allows us to calculate the significance of the observed discrimination, while the second one (STAT-ITCF, Gouet et al. 1985) gives the percentage of points which are not classified with those of their own species.

2 Results

2.1 Comparative Analysis of the Acoustic Structure of Calls

Warning Calls. Two types of warning calls are present in both species; the first belongs to the low-pitched register, the second to the high-pitched register. Only the latter will be exemplified here.

In both species, high-pitched warning calls are mainly given by adult females, juveniles, and subadults of both sexes. Either unitary, binary, or ternary chirps can be heard (Fig. 1). However, unitary calls are most frequent in *C. nictitans*, while *C. cephus* mainly gives either rhythmic sequences of unitary calls or binary calls. Ternary chirps are seldom used by either species. Comparative measurements of the mean values of four parameters show significant differences ($p < 0.05$; Table 1).

The correspondence analysis does not totally distinguish the two species (Fig. 2) and an overlapping between them is obvious on the factorial plane (71% of the total inertia). While the majority of *cephus* points are localized in north-east of the plane, some are mixed with *nictitans* points in the three other quadrats. The slope of the fundamental tone (horizontal axis: 44.5% of the relative inertia) and the call duration (vertical axis: 47.5% of the relative inertia) seem to contribute best in separating the two species.

Discriminant analyses show that the separation between the two species is significant ($p < 0.001$) and mainly relies on the call duration. The classification accuracy reaches 82%.

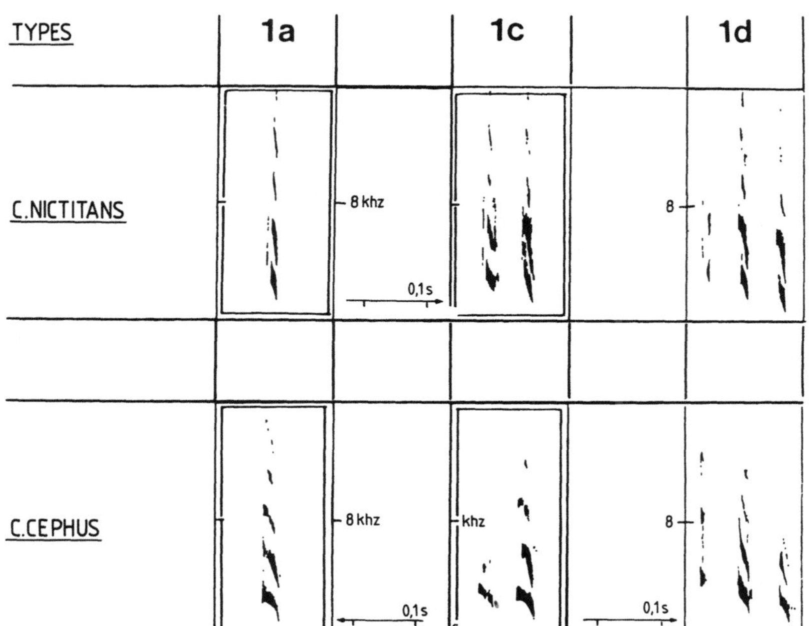

Fig. 1. Spectrographic analyses of unitary, binary, and ternary chirps (*1a, 1c,* and *1d*, respectively) given by *C. nictitans* and *C. cephus*

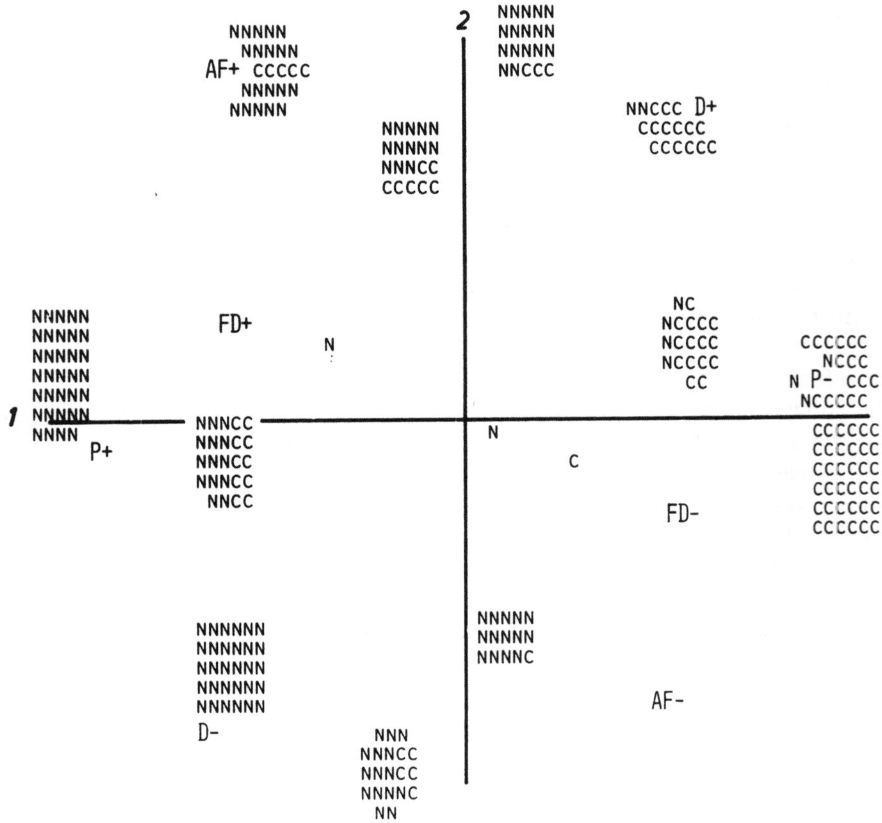

Fig. 2. Correspondence analysis of 298 unitary warning chirps (type 1a) of *C. nictitans* (N = 171) and *C. cephus* (C = 127) species based on the four following parameters: *D* = duration; *FD* = dominant frequency; *AF* = amplitude frequency; *P* = slope of the fundamental tone. The signs "+" and "–" denote the two classes of disjunctive variables in which each parametric sample has been subdivided (see the text for interpretation)

Table 1. Comparative measurements of acoustic parameters of warning chirps (type 1) given by *C. nictitans* and *C. cephus*

Parameters	*C. nictitans* [a]			*C. cephus* [a]			t-Test
	$\bar{X}$	CV	n	$\bar{X}$	CV	n	
Duration (ms)	17	33	114	24	24	45	< 0.05
Amplitude frequency (kHz)	1.94	31	114	1.6	33	124	< 0.05
Dominant frequency (kHz)	3.40	15	114	2.9	12	168	< 0.05
Slope (kHz s^{-1})	121	28	114	79	44	123	< 0.05

[a] $\bar{X}$ = mean value; *CV* = coefficient of variation; *n* = number of calls analyzed.

System of Cohesion Calls. Three main vocal types are used by *C. nictitans* and *C. cephus* to maintain troop cohesion and interindividual relationships: (1) a high-pitched call, named type 6 contact call (Gautier 1975, 1978) which is mainly used in close exchanges between young and their mothers; (2) a low-pitched call, named type 2 cohesion call which frequently anticipates and accompanies troop movements. They are given in sequences by adult females and immatures and spread throughout the troop by phonoresponses among individuals who most often are not in view of another. In mixed troops, such phonoresponses can be either intra- or interspecific; (3) A medium-pitched call, named type 3 lost call. It is mainly given by young infants when they are away from their mothers and by immature individuals who have lost contact with their troop. This can serve to reunite a mother with her infant.

Type 6 Contact Calls. These calls vary greatly in duration, loudness, and energy localization from a weak, practically nonvoiced sound (type 6′, Fig. 3) to the harsh and loud type 6–7, the latter being able to change into a distress call. Among them, type 6 calls are the most frequently heard.

In both species, they have a moderate intensity, a narrow bandwidth of fundamental tone, a decreasing frequency modulation, and a quavering. However, comparative measurements of the mean values of six acoustic parameters show significant differences in duration, dominant frequency, upper frequency and, most obviously, the quavering rhythm (Table 2).

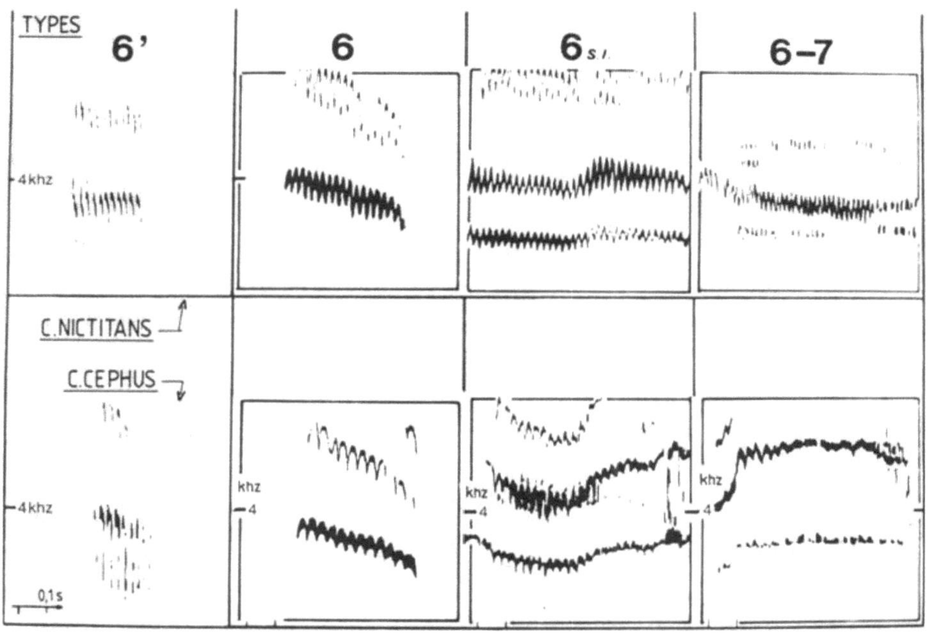

Fig. 3. Spectrographic analyses of the graded series of contact trills (from type 6′ to type 6–7) given by *C. nictitans* and *C. cephus*. (*6 s.i.* = strong intensity)

Table 2. Comparative measurements of acoustic parameters of contact trills (type 6) given by *C. nictitans* and *C. cephus* (abbreviations as in Table 1)

Parameters	*C. nictitans* (n = 22)		*C. cephus* (n = 92)		t-Test
	$\bar{X}$	CV	$\bar{X}$	CV	
Duration (ms)	385	43	560	51	< 0.01
Upper frequency (kHz)	3.8	11.5	4.03	13	< 0.05
Lower frequency (kHz)	2.1	21	2.40	29	= 0.05
Dominant frequency (kHz)	3	16	3.30	17	< 0.05
Amplitude frequency (kHz)	1	24	0.93	41	> 0.05
Rhythm of quavering (Hz)	46	5.5	36	17	< 0.01

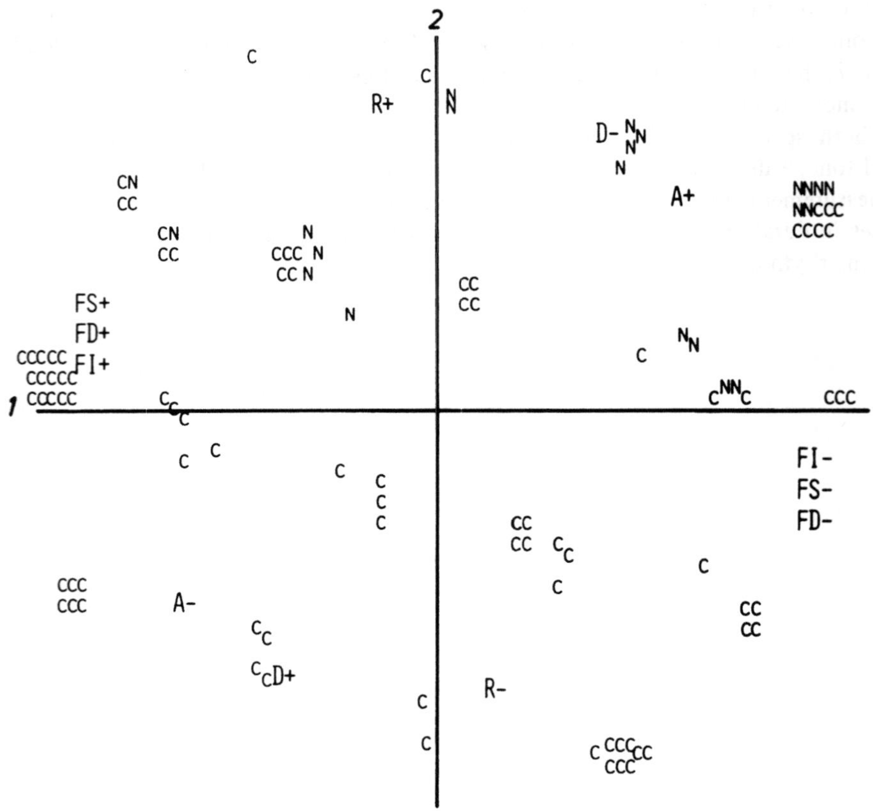

Fig. 4. Correspondence analysis of 108 contact trills (type 6) of *C. nictitans* (N = 22) and *C. cephus* (C = 86) based on the following six parameters: *D* = duration; *FD* = dominant frequency; *FS* = superior frequency; *FI* = lower frequency; *A* = frequency amplitude of the quavering; *R* = rhythm of the quavering. The signs "+" and "–" denote the two classes of disjunctive variables in which each parametric sample has been subdivided (see text for interpretation)

Correspondence analysis separates *nictitans* and *cephus* species (Fig. 4) which appear localized in the north-east and in the south-west of the factorial plane (61% of the total inertia), respectively. The rhythm, the duration and, to a lesser degree, the amplitude frequency contribute best in separating the species (40, 34 and 19% of the total relative inertia along the vertical axis, respectively).

Discriminant analyses reveal that the separation between the two species is significant ($p < 0.001$) and mainly relies on the rhythm; the classification accuracy reaches 90%.

Type 2 Cohesion Calls. These calls belong to a graded vocal system which begins with huddling calls of the young (type 2′, Fig. 5) and ends with lost calls (type 3).

Among them, different call types occur, the most commonly heard being type 2. The latter are very similar in the two species and look like low-pitched discontinuous trills. Their mean duration does not differ between species while all other parameters analyzed differ significantly (Table 3).

Correspondence analysis shows that the two species are clearly separated by the horizontal axis of the factorial plane (70% of the total inertia, Fig. 6). The pitch, the rhythm and, at lower levels, the dominant and the superior frequencies are the parameters which contribute best in separating species along the vertical axis (30, 29, 20.5, and 20.5% of the relative inertia, respectively).

According to discriminant analyses, the discrimination is significant ($p < 0.001$) and is mainly due to pitch and rhythm. The classification accuracy reaches 100%.

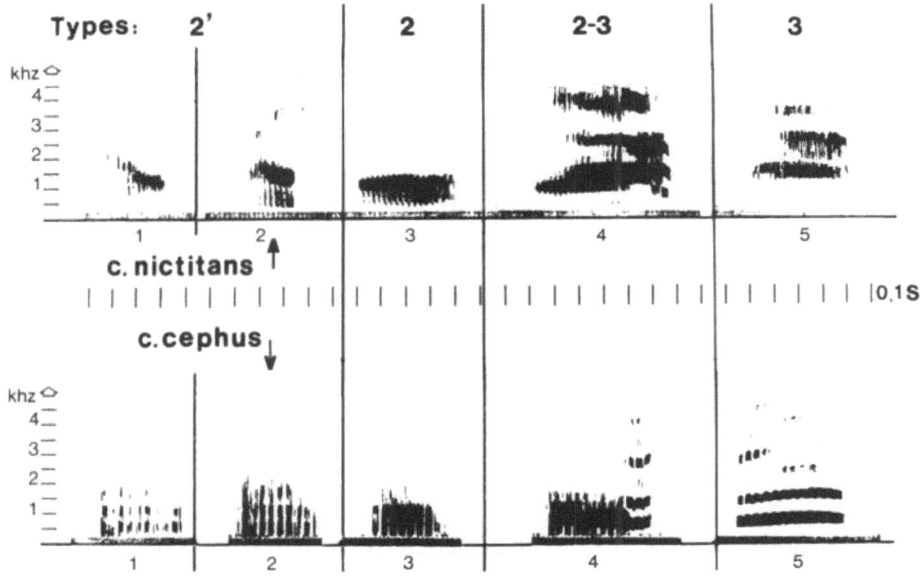

Fig. 5. Spectrographic analyses of the graded series of cohesion calls (from *type 2′* to *type 3*) given by *C. nictitans* and *C. cephus*

Table 3. Comparative measurements of acoustic parameters of cohesion trills (type 2) given by *C. nictitans* and *C. cephus* (abbreviations as in Table 1)

Parameters	*C. nictitans* (n = 41)		*C. cephus* (n = 33)		t-Test
	$\bar{X}$	CV	$\bar{X}$	CV	
Duration (ms)	122	14.5	131	22	> 0.05
Upper frequency (kHz)	0.96	15.4	1.17	16	< 0.01
Dominant frequency (Hz)	626	21	746	15	< 0.01
Pitch (Hz)	236	10.4	155	17	< 0.01
Rhythm of quavering (Hz)	47.4	11	31.2	16.5	< 0.01

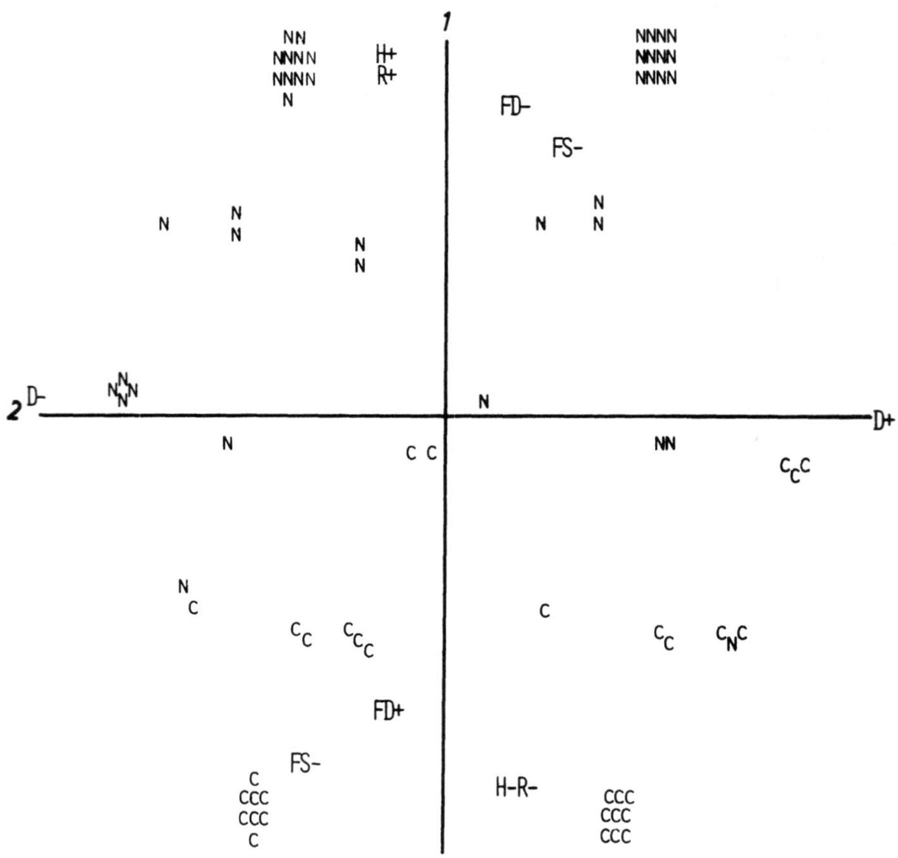

Fig. 6. Correspondence analysis of 74 cohesion calls (type 2) of *C. nictitans* (N = 41) and *C. cephus* (C = 33) based on the following five parameters: *D* = duration; *FD* = dominant frequency; *FS* = superior (upper) frequency; *H* = pitch; *R* = rhythm of the quavering. The signs "+" and "–" denote the two classes of disjunctive variables in which each parametric sample has been subdivided (see the text for interpretation)

Type 3 Lost Calls. Lost calls of *C. nictitans* and *C. cephus* are easier to identify than contact or cohesion calls: *nictitans* calls are noisy while those of *cephus* show a tonal structure (Fig. 5). They both have a longer duration and louder intensity than all calls of the system. Young vocalizers appear very excited and their calls display a high degree of inter- and intra-individual variability.

Table 4. Comparative measurements of acoustic parameters of lost trills (type 3) given by *C. nictitans* and *C. cephus* (abbreviations as in Table 1)

Parameters	*C. nictitans*			*C. cephus*			t-Test
	$\bar{X}$	CV	n	$\bar{X}$	CV	n	
Duration (ms)	410	26	29	329	22	85	< 0.01
Upper frequency (kHz)	3.9	37	29	4.8	30	85	< 0.01
Dominant frequency (kHz)	1.4	36	29	1.4	27	85	> 0.05
Pitch (Hz)	364	41	29	950	32	85	< 0.01
Rhythm of quavering (Hz)	45	12.3	20	33	12	61	< 0.01

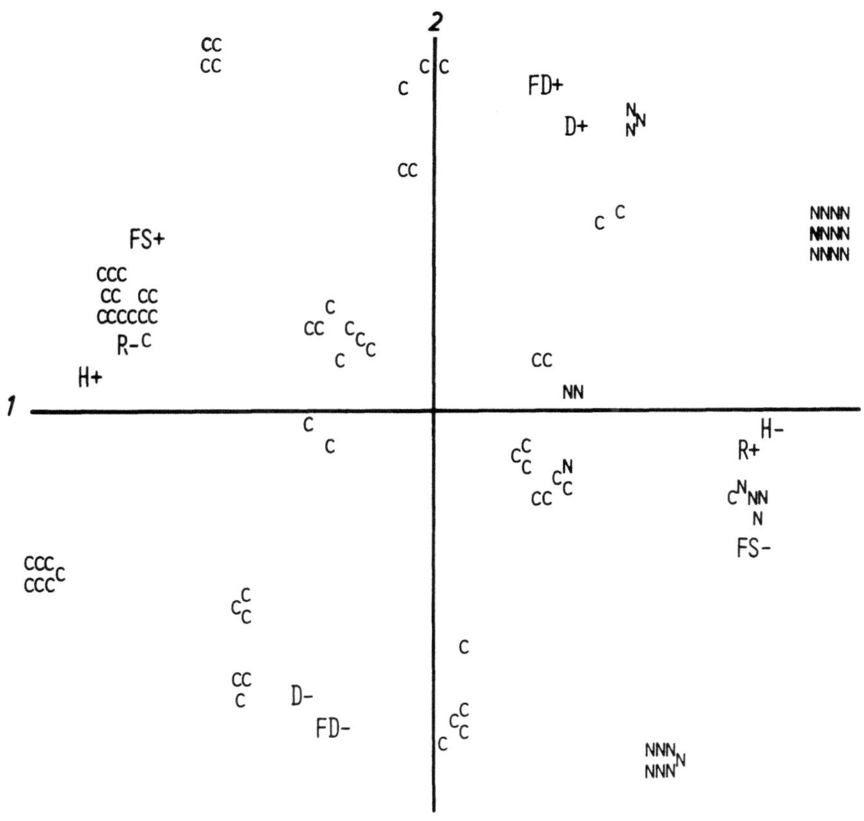

Fig. 7. Correspondence analysis of 91 lost trills (type 3) of *C. nictitans* (N = 29) and *C. cephus* (C = 62) based on the same five parameters as those of Fig. 6

Four of five acoustic parameters which characterize these calls differ between species (Table 4). On the factorial plane (58% of the total inertia) of the correspondence analysis, species are differently localized (Fig. 7), according to the pitch, the rhythm, and the upper frequency of their calls (35, 20, and 27% of the relative inertia, respectively).

Discriminant analyses confirm that this discrimination is significant ($p < 0.001$) and mainly relies on pitch and rhythm; the classification accuracy reaches 96.7%.

2.2 The Rhythm of Call Quavering

Considering the cohesion-contact vocal system analyzed, we observe that for every species the three calls possess a very similar mean value of quavering, around 37 Hz for *C. cephus* and 46 Hz for *C. nictitans*, while interspecific differences are significant (Fig. 8).

Sex-Related Differences. Sixty-one calls recorded from four male and four female *C. cephus* less than one-year-old have been analyzed. Results show a large overlap of values from both sexes (Table 5).

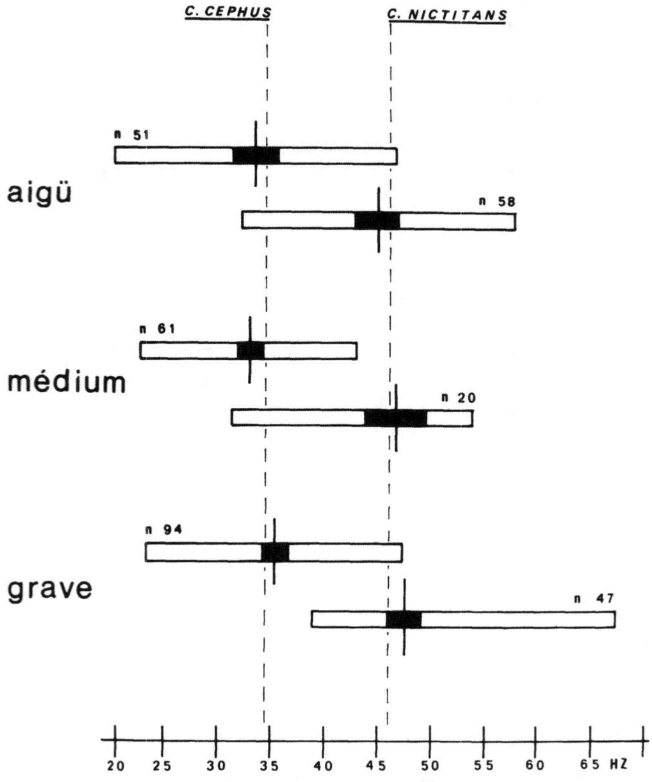

Fig. 8. Mean values (*vertical black lines*), confidence limits of mean ($p = 95\%$; *black bars*), and range values (*open bars*) of the rhythm of the quavering (in Hz) for the three calls of the cohesion/contact system (high-, medium-and low-pitched sounds for "aigu", "medium" and "grave", respectively) given by *C. cephus* and *C. nictitans*. n = Size of samples for each vocal type

Table 5. Rhythm values of the quavering for lost trills (type 3) given by four male and four female *C. cephus* less than 1-year-old

Individuals	1			2			3			4		
Rhythm of quavering	$\bar{X}$	CV	n	$\bar{X}$	CV	n	$\bar{X}$	CV	n	$\bar{X}$	CV	n
Males	28	0.15	2	28	8	17	33	7	21	36	12.4	3
Females	31	0	2	32	2.6	8	36	0	2	40	3.5	6

Age-Related Differences. No correlation is found between age and rhythm values for five young monkeys from 2- to 42-months-old. Either high or low values of the quavering rhythm may be observed in young and old animals (Table 6).

In fact, type 2 cohesion calls of two *C. cephus* of the same age and sex may possess the same duration and the same pitch, while differing most significantly in their rhythm values (Table 7).

Individual-Related Differences. Considering the rhythm values of the three call types given by two *C. cephus*, it appears that for each animal a similar value characterizes the whole call system and, moreover, shows a low coefficient of variation (Table 8).

Furthermore, for a sample of type 2 calls given by the same monkey at the same age, it appears that the coefficient of variation of rhythm is far lower than that of all other acoustic parameters (Table 9), showing that this parameter is the least subject to intraindividual variations induced by a situational and/or emotional context.

Table 6. Rhythm values of the quavering and their coefficients of variation for contact trills (type 6) given by six *C. cephus* monkeys at different ages

Age in months	2	6	18	18	42
Rhythm of quavering (Hz)	29	42	29	44	37
Coefficient of variation	4.5	3.4	6.4	2.0	4.4
Number of calls	13	22	18	7	21

Table 7. Comparative measurements of acoustic parameters of cohesion trills (type 2) given by two *C. cephus* monkeys of the same age and sex

Parameters	Monkey A			Monkey B			t-Test
	$\bar{X}$	CV	n	$\bar{X}$	CV	n	
Duration (ms)	142	33	27	146.4	25	25	> 0.05
Dominant frequency (kHz)	1.25	30	24	0.93	9	9	< 0.05
Pitch (kHz)	0.18	12	25	0.17	13	24	> 0.05
Rhythm of quavering (Hz)	43.3	4	27	32	7	24	< 0.01

Table 8. Comparative measurements of the rhythm value of the quavering for the three calls of the cohesion-contact system given by two *C. cephus* monkeys

Call types	Type 2			Type 3			Type 6		
	$\bar{X}$	CV	n	$\bar{X}$	CV	n	$\bar{X}$	CV	n
Monkey 1	28	5.4	3	28	0.15	2	29	4.5	13
Monkey 2	32	5.6	14	31	0	3	29	6.4	16

Table 9. Acoustic parameters and their coefficients of variation for cohesion trills (type 2) given by the same *C. cephus* monkey at the same age

Parameters	Type 2 cohesion trills		
	$\bar{X}$	CV	n
Duration (ms)	142	33	27
Dominant frequency (kHz)	1.25	30	24
Pitch (kHz)	0.18	12	25
Rhythm of quavering (Hz)	43.3	4	27

Table 10. Comparative measurements of pitch and rhythm of quavering for cohesion trills (type 2), given by three monkeys at different ages

Individuals	Before puberty			Puberty			After puberty			t-Test
	$\bar{X}$	CV	n	$\bar{X}$	CV	n	$\bar{X}$	CV	n	
C. cephus ♀		(3 y.o)						(9 y.o)		
Pitch, Hz	155	10	15				122	5	18	< 0.01
Rhythm, Hz	41	11	16				38	9	21	> 0.05
C. nictitans ♂		(1.5 y.o)			(6.5 y.o)					
Pitch, Hz	231	5	7	168	4.5	7				< 0.01
Rhythm, Hz	52	7.5	12	40	6	11				> 0.05
C. nictitans ♂		(4 y.o)						(18 y.o)		
Pitch, Hz	206	7	6				144	--	1	–
Rhythm, Hz	31	3	7				29	–	1	–

Maturational Processes. We have previously shown (Gautier and Gautier-Hion 1977) that in guenons there are maturational processes in call structure reflected in regular decrease of call pitch with aging which is inversely related to body weight. In males such a decrease of call pitch undergoes an abrupt inflexion during puberty when a corresponding increase in growth rate occurs. This corresponds to breaking of the male voice (Table 10).

If we compare the mean values of pitch and rhythm for cohesion calls given by the same female *C. cephus*, 1 year before sexual maturity and 5 years later, it appears that the female's voice loses 20% of its value during this 6-year interval, while over the same time period the quavering value of its cohesion calls is not significantly modified (Table 9). Similar results characterize cohesion calls given by *C. nictitans*: a male, recorded when he was 1.5-years-old and 5 years later when he reached puberty, showed a 30% decrease of call pitch, while the rhythm of call quavering remained approximately unchanged. After 14 years, the pitch of cohesion calls of another male also decreased by 30% and the rhythm by less than 1%.

Thus, in calls with a role in intratroop cohesion and interindividual relationships the rhythm of quavering has a number of characteristics which suggest it might operate in species and individual recognition: (1) it possesses a similar mean value for the whole call system, both at the species and at the individual level; (2) it does not appear to be sex-dependent; (3) it is not age-related; (4) it does not undergo any maturational change; (5) it does not suffer significantly from the situational context; and (6) it has one of the lowest coefficients of variation.

It remains, however, to demonstrate whether monkeys are or are not actually able to rely on such a parameter.

3 Discussion

Among the calls studied above, alarm chirps show the least species specificity. Similar results have frequently been stressed for a number of species, e.g. both birds and monkeys, and are mainly explained as a case of parallelism and by the fact that such calls have undergone little selective pressure toward species specificity (e.g., Marler 1957). In fact, we observed that the alarm calls of a bird *(Tricophorus calurus)*, living sympatrically with the two guenons studied in Gabon, also include a warning chirp that is difficult to distinguish from those of monkeys (Gautier 1975). Such convergences between different taxa have been discussed by Maier et al. (1983).

It could be expected that calls playing a role in interindividual relationships require more specificity (see, e.g., Snowdon et al. 1983). Analyses show that, despite the interspecific structural similarities in the three studied calls of this system, there are significant interspecific differences in the values of a number of parameters. Considering the whole system of cohesion-contact calls, the rhythm of quavering is a discriminant parameter for all call types, and one of the most discriminant parameters, for all of them. Consequently, it can be regarded as the most species-specific call feature. This needs further investigation.

Two main trends are found in the literature relevant to species specificity and individual recognition in calls. Firstly, individual recognition relies on a combination of all acoustic features: this is the case, for example, for *isolation peeps* of squirrel monkeys (Symmes et al. 1979) or of *Lemur catta* (Macedonia 1986); secondly, one call parameter is found to be more discriminant than others, this is the case of the time interval between the two masts of the *chuck* calls of squirrel monkeys (Smith et al. 1982).

Whatever the case may be, conditions under which calls are usually given have to be taken into account to understand why one acoustic feature has been selected over others. Constraining conditions which may alter the suitability of calls for species or individual identification are of at least three types. They may result from the complex vocal atmosphere provided by the richness of surrounding noises, as e.g. in nesting sites of colonial birds (Beer 1970). In such situations, the temporal patterning of calls has been suggested to play a major role in individual recognition, as in gannets (White et al. 1970) or in penguins (Jouventin 1972, 1979).

Another main constraint results from differential alteration of acoustic parameters through the environment. Thus, Waser (1977) has shown in forest mangabeys that while whoop gobbles of different adult males can be easily discriminated near the source on the basis of their tonal structure, only their temporal patterning remains conspicuous at a distance, whereas other acoustic parameters are considerably altered.

Thus, it is not surprising to find an elaborate temporal patterning for most loud calls of many forest monkeys (see, e.g., Gautier and Gautier-Hion 1977; Quris 1980) and to observe, for example, that several species share the same "boom" loud call whose temporal sequences of emission are the only criteria on which species identification may rely. Such is the case, for example, for *Cercopithecus pogonias* and *C. neglectus*, two sympatric guenons (Gautier 1975).

Maturational constraints must also be considered. Thus, in *Saimiri sciureus*, Lieblich et al. (1980) have shown that the structure of *isolation peeps* changes with age until 36 weeks as the call develops into its adult form. In guenons, the enlargement and other developmental factors of the vocal tract (larynx and vocal annexes) observed during ontogeny (Gautier 1971) have been found to induce important changes in frequency components of calls as well as in their overall tone (Gautier and Gautier-Hion 1977). Such progressively changing call characteristics may constitute a perceptual basis for age class recognition and are therefore poor candidates for markers of individual identity. On the contrary, temporal patterns and especially the rhythm of call quavering, which probably depend directly on neural control, are not modified by aging. Consequently, they appear as the most suitable parameters for individual recognition and it is not surprising to note that temporal patterns have been suggested to play such a role in a number of cases, in different taxa.

Cohesion calls of several guenon species (about 17 of 27) are characterized by a rhythmic frequency modulation (Gautier 1988). Each species is characterized by a well-defined, species-specific value of the quavering rhythm of cohesion calls, ranging from a minimum of 18 Hz (in *C. aethiops*) to a maximum of 50 Hz *(Allenopithecus nigroviridis)*. Such a gradient is not correlated with species body size and it can be hypothesized that selective pressure has operated on this physical parameter as a cue for reproductive isolation. The question remains, however, whether and how species whose cohesion calls are not quavered (such as species of the *mona* group) realize such recognition and why they have developed a system different from that of the majority of guenons.

In summary, one can hypothesize different levels of call perception in forest guenons for which visual cues are frequently lacking. The whole vocal profile of calls would allow species identity; the pitch and frequency component of calls which regularly change with aging would permit an approximate age and sex recognition; more subtle variations of the above components and of the overall call duration would be relevant to internal mood and situational context. Finally, at least for cohesion calls given without the support of visual signals, the rhythm of calls' quavering would be the most probable cue for individual recognition.

4 Summary

The two sympatric guenons, *C. nictitans* and *C. cephus*, possess very similar vocal repertoires, both at the structural and functional levels. However, while alarm calls of both species are imperfectly discriminated, cohesion and contact calls possess significant differences for the values of a number of acoustic parameters. These include especially the rhythm of quavering and the pitch, which may contribute to species recognition.

Furthermore, the rhythm of quavering does not undergo age-related changes or situational context-related variations, and appears to be the cue best fitted for individual recognition.

Acknowledgments. We are grateful to Dr. Todt and his co-workers for their invitation to participate in the stimulating symposium on primate communication held in Berlin in 1986. We thank Mr. R. Quris for his assistance in using computer programs and statistical analyses.

References

Bachacou J, Masson JP, Millier C (1981) Manuel de la programmathèque statistique. AMANCE 81, CNRF, INRA

Beer CG (1970) Individual recognition of voice in the social behaviour of birds. Adv Stud Behav 3: 22–74

Benzécri JP (1973) L'analyse des données. T2: l'analyse des correspondances. Dunod, Paris

Dutrillaux B, Couturier J, Muleris M, Chauvier G (1982) Chromosomal phylogeny of forty-two species or subspecies of cercopithecoids (*Primates, Catarrhini*). Ann Genet 25:96–109

Fénelon JP (1981) Qu'est-ce-que l'analyse de données? Lefonen, Paris

Gautier JP (1971) Etude morphologique et fonctionnelle des annexes extralaryngées des Cercopithecinae: liaison avec les cris d'espacement. Biol Gabonica 7:229–267

Gautier JP (1975) Etude comparée des systèmes d'intercommunication sonore chez quelques cercopithecines forestiers africains. Thèse Doct, Univ Rennes

Gautier JP (1978) Répertoire sonore de *Cercopithecus cephus*. Z Tierpsychol 46:113–169

Gautier JP (1988) Interspecific affinities as deduced from vocalizations among African guenons. In: Gautier-Hion A, Bourlière F, Gautier J-P, Kingdon J (eds) A primate radiation: evolutionary biology of the African guenons. Cambridge Univ Press, Cambridge, pp 194–226

Gautier JP, Gautier-Hion A (1969) Les associations polyspecifiques chez les cercopithecidae du Gabon. Terre Vie 2:104–201

Gautier JP, Gautier-Hion A (1977) Communication in Old World monkeys. In: Sebeok TE (ed) How animal communicates. Indiana Univ Press, Bloomington, pp 890–964

Gautier-Hion A, Gautier JP (1974) Les associations polyspécifiques de cercopithèques du plateau de M'passa, Gabon. Folia Primatol 22:134–177

Gouet H, Phillipeau G, Tranchefort J, Verneau M (1985) Manuel d'utilisation de STAT-ITCF. Serv Et Stat, Stn Exp ITCF, Boigneville

Jouventin P (1979) Le chant du manchot adélien *(Pygoscelis adelia)*: rôle dans la reconnaissance 43:175–185

Jouventin P (1979) Le chant du manchot adelien (*Pygoscelis adelia)*: rôle dans la reconnaissance individuelle et comparaison avec le manchot empereur non territorial. Oiseau 49:31–37

Lieblich AK, Symmes D, Newman JD, Shapiro M (1980) Development of isolation peep in laboratory-bred squirrel monkeys. Anim Behav 281:1–9

Macedonia JM (1986) Individuality in a contact call of the ring-tailed lemur *(Lemur catta)*. Am J Primatol 11:163–179

Maier V, Rasa PA, Scheich H (1983) Call-system similarity in a ground-living social bird and a mammal in a bush habitat. Behav Ecol Sociobiol 12:5–9

Marler P (1957) Specific distinctiveness in the communication signals of birds. Behaviour 11:13–39

Marler P (1973) A comparison of vocalizations of red-tailed monkeys and blue monkeys (*Cercopithecus ascanius* and *C. mitis*) in Uganda. Z Tierpsychol 33:223–247

Quris R (1980) Emission vocale de forte intensité chez *Cercocebus galeritus agilis*: structure, caractéristiques spécifiques et individuelles, modes d'émission. Mammalia 44:35–50

Ruvolo M (1988) Genetic evolution in the African guenons. In: Gautier-Hion A, Bourlière F, Gautier J-P, Kingdon J (eds) A primate radiation: evolutionary biology of the African guenons. Cambridge Univ Press, Cambridge, pp 127–139

Smith HJ, Newman JD, Hoffman H, Fetterly K (1982) Statistical discrimination among vocalizations of individual squirrel monkeys *(Saimiri sciureus)*. Folia Primatol 37:267–279

Snowdon CT, Cleveland J, French JA (1983) Responses to context and individual specific cues in cotton tamarin long calls. Anim Behav 31:92–101

Sokal RR, Rohlf FJ (1969) Biometry: the principles and practice of statistics in biological research. Freeman, San Francisco

Symmes D, Newman JD, Talmage-Riggs G, Lieblich AK (1979) Individuality and stability of isolation peeps in squirrel monkeys. Anim Behav 27:1142–1152

Waser P (1977) Individual recognition, intragroup cohesion and intergroup spacing evidence from sound playback to forest monkeys. Behaviour 60:28–74

White SJ, White REC, Thorpe WH (1970) The acoustic basis for individual recognition by voice in the gannet. Nature (London) 225:1156–1158

An Analysis of Toque Macaque Cohesion Calls from an Ecological Perspective

WOLFGANG DITTUS[1]

The manner in which individuals in social groups space themselves in relation to each other and to other groups partly reflects how they might share limiting resources with one another, select their mates, care for their young, and protect one another against predators or resource competitors. Within most primate societies, not all members are equally privileged in using the groups' available resources, be it food, water, mates, or refuges. Dominance hierarchies, determined by the social histories among group members, reflect which families, peer groups or similar cliques, and which individuals within such subgroupings, are more privileged than others. Competition and cooperation among group members are especially strongly expressed in relation to the exploitation and sharing of food resources. Hence, an individual's choice of social partners, especially during foraging, may strongly influence its chances of surviving and reproducing (Dittus 1977a, 1986).

Much of what primates communicate about concerns the exercise of their particular "rights" in society. To this end, a number of threat and appeasement postures and vocalizations serve as communicative tools to manipulate social partners, particularly at times of conflict.

There exists yet another class of equally important signals that guides individuals in their more or less chronic selections of nearest neighbors. Monkeys, particularly macaques, differ markedly in their individual morphologies and these in themselves act as visual guides. One aim of this chapter is to examine the vocal signals that assist group members in selecting, and in staying close to, their partners in daily social life. These signals concern the structurally and contextually diverse vocalizations that generally have been considered together as types of *cohesion calls, contact calls,* or *coo calls* in a variety of Old World primates (Itani 1963; Lindburg 1971; Green 1975; Struhsaker 1967; Cheney and Seyfarth 1982; Caldecott 1986; Mitani 1986).

I propose that this class of calls constitutes at least three kinds of related signals that might be broadly classified as *contact calls, lost calls,* and *food calls* (Table 1). They share the role of guiding individuals in their selection of social partners. They differ in their additional specialized functions, as will be examined in this chapter.

1 The Smithsonian Institution, Washington, DC, USA and Institute of Fundamental Studies, Kandy, Sri Lanka

Todt/Goedeking/Symmes (Eds.)
Primate Vocal Communication

Table 1. Physical characteristics of different cohesion call types

Call type	Frequency range (Hz)	Frequency emphasis (Hz)	Duration (s)	Loudness	Sound quality	Pitch trajectory	Figure
Contact calls							
Hum (infant)	0.8–1.0		0.10–0.50	Very quiet	Pure tone	Flat	1A
Hum (adult)	0.3–0.5		0.10–0.50	Very quiet	Pure tone	Variable	1B
Grunt	0.1–1.0		0.10–0.50	Very quiet	Atonal	Flat, broad spectrum	2A
Food call	0.5–4.5	2.5–4.5	0.20–0.50	Loud	Tonal squeal	Sharp rise	3
Lost calls							
Lost-1	0.4–0.8	0.6–0.8	0.40–0.75	Moderately loud	Pure tonal moan	Slight rise and fall	4B
Lost-2	0.5–1.8	>1.0	0.30–0.60	Loud	Tonal yell	Sharp rise and fall	4C
Lost-3							
part 1	0.5–3.0		0.15	Loud	Tonal whistle	Sharp rise	1D
part 2	1.8–0.5	>1.0	0.20–0.35	Loud	Tonal yell	Slight fall	1E
Lost-4							
part 1	1.8–3.5	3.5	0.13–0.16	Very loud	Piercing	Sharp rise	4F
part 2	1.7–0.6	1.7	0.15–0.20	Very loud	Yell	Slight fall	
Whistle							
(adult)	0.5–>5.0	>5.0	0.15	Very loud	Tonal whistle	Sharp rise	4G
(infant)	1.0–>5.0	>5.0	0.20–0.25	Loud	Tonal whistle	Sharp rise	4H
Infant separation calls							
Sep-1	0.70–0.90	Nil	0.8–1.10	Very quiet	Pure tone	Flat	5A
Sep-2	0.70–1.0	1.0	0.8–1.10	Quiet	Tonal moan	Slight rise and fall	5B
Grunt	0.20–2.0	Nil	0.30–0.50	Very quiet	Atonal	Flat, broad spectrum	6

1 Methods

1.1 The Study Area, Period, and Subjects

The behavior, ecology, and demography of toque macaques *(Macaca sinica)* were studied at Polonnaruwa, Sri Lanka, for nearly two decades. Observations were continuous from 1968 to 1972, and 1977 to 1987, and intermittent between 1972 and 1977.

The natural dry evergreen forest, which the macaques inhabit at the study site, consists of a dense shrub layer from 2-5 m, and is overlain by trees whose crowns make up a more or less continuous canopy layer between 8 and 18 m. A discontinuous layer of emergent trees reaches up to 30 m (Dittus 1977b, 1985). In disturbed areas of the site, the shrub layer has been removed so that a rich herb layer and tall trees result in a parkland setting. Toque macaques are forest-bound, tending to avoid large open areas. Visibility within the forest is limited.

The study population consists of 25 social groups of approximately 500 individuals. All macaques in the population were individually identified (Dittus and Thorington 1981). Behaviors concerning food and lost calls were recorded whenever these vocalizations were heard. The macaques were well habituated and could be approached to within 1-3 m. Most observations were made within 3-15 m.

1.2 Sound Recording and Analysis

The vocalizations considered in this report are distinctive and readily identified by ear. Occurrences of *contact, food,* and *lost calls* were noted mostly while sampling ecological and behavioral data from focal animals. Whenever possible, I recorded the identities and behaviors of the vocalizing individuals and of the animals responding to the calls.

Examples of vocalizations were recorded onto magnetic tape. An Uher tape recorder, model IC at tape speed 19 cm s^{-1} was used in conjunction with an Uher microphone with a plastic windscreen. Recordings were made with the microphone at 0.5-30 m from the vocalizing animal. Most recordings of vocalizations were transferred onto cassette tape (BASF chromium dioxide super II) using a Sony Professional Walkman, model WM-D6C, tape recorder. Spectrographic tracings were made from these cassette recordings using a Kay elemetrics Sonagraph model 7029A set at the wide-band filter, the FL1 shaping selector, and the 80 to 8000 Hz frequency scale. For each call type, I selected two to ten spectrographs for measurements of frequency range and temporal patterns.

2 Results

2.1 The Physical Structure (Syntax) of Call Types

Hum Contact Calls. There were two main types of low amplitude contact calls, labeled *hums* and *grunts* that, under normal field conditions, were audible for only a few meters by the human observer.

Hums varied considerably in the trajectory of pitch, harmonic structures, duration, and amplitude modulation. For example, hums rose, fell, or rose and fell in pitch, and

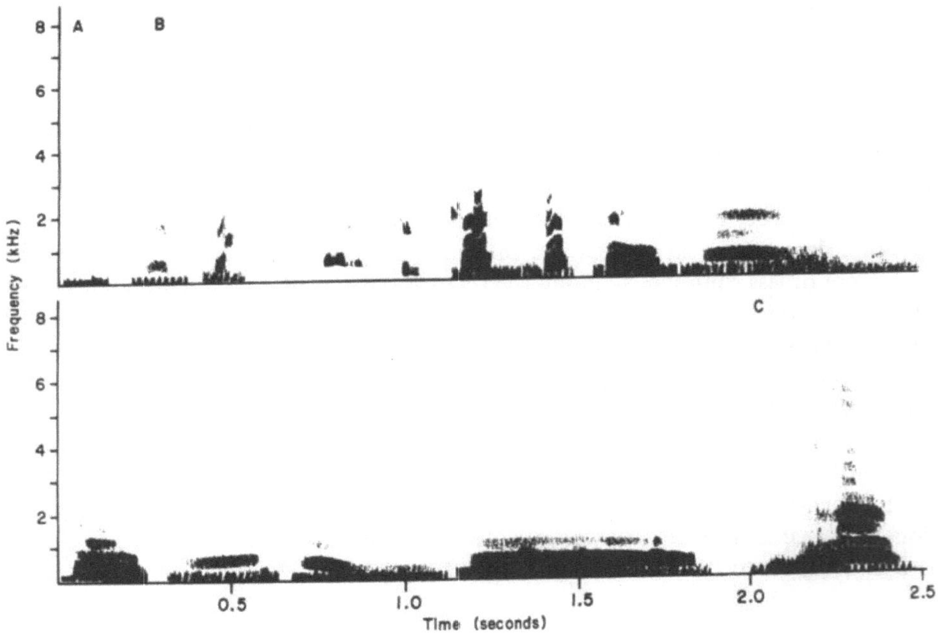

Fig. 1. Spectrographs of hum contact calls of toque macaques. The hum of a young juvenile (*A*) is of a higher frequency than the many structurally diverse hums produced by adults (*B*), some of which are combined with a grunt (*C*)

the sound energy was variably modulated within a call (Fig. 1B). Thus, depending on one's classificatory criteria, many hum types might be recognized. Hums also were combined with other hums or grunts (Fig. 1C) into short variable sequences. Hums of infants and juveniles (Fig. 1A) were higher pitched than those of adults (Fig. 1B).

Grunt Contact Calls. Compared to hums, grunts spanned a broader frequency range, sounded less pure-toned, and were lower pitched. Grunts, too, were modulated in duration, amplitude, and frequency. Adult grunts (Fig. 2B) appeared to be ontogenetically derived from higher pitched infant grunts (Fig. 2A).

Food Calls. A typical food call was a loud, high-pitched, and drawn-out "whee" sound which normally began with a brief inflection in pitch. Some food calls were heard from at least 100 m by humans, and by the macaques as well, judging from their responses.

Hum vocalizations often preceded and graded into food calls. Structurally, food calls appeared to be derived from hum calls through a sudden burst of intensity accompanied by an increase in pitch and duration (Fig. 3). Food calls were acoustically distinct as a type.

Food calls normally were interspersed with grunt and hum contact calls, and like them, were highly contagious. Hence, food calls were most often heard in brief choruses of intense humming and grunting accentuated with several loud squeaky food calls.

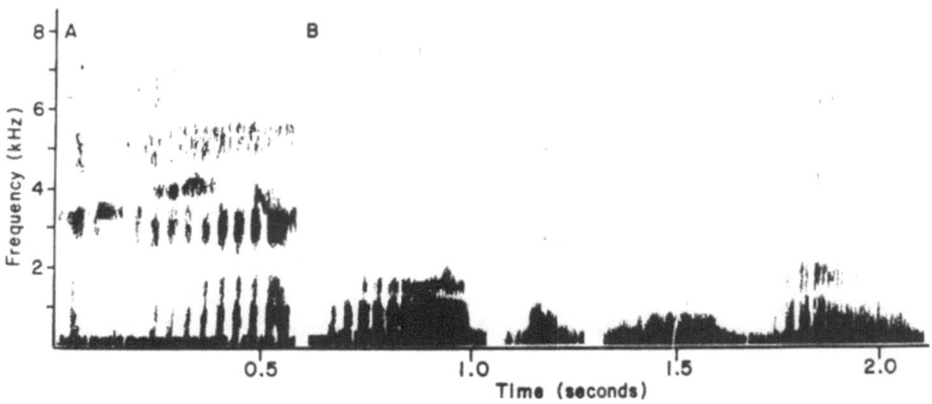

Fig. 2. Spectrographs illustrating the structural variation of the grunts of toque macaques. The grunt of a young juvenile (*A*) is of a higher pitch than the grunts of older animals (*B*)

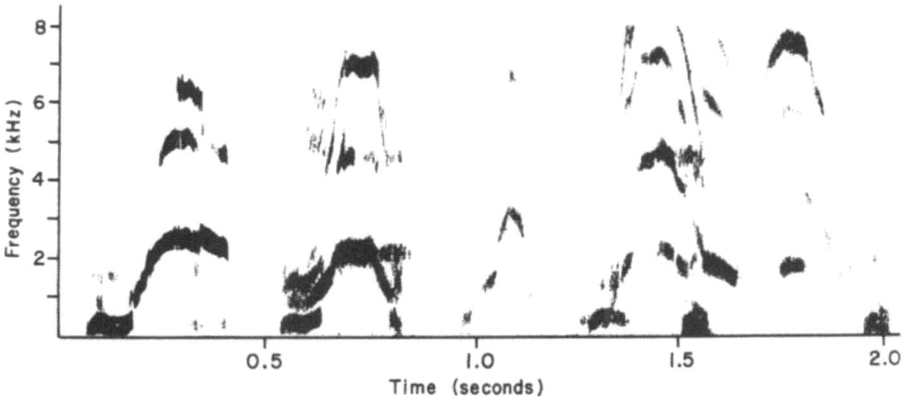

Fig. 3. Spectrographs of five different food calls illustrating their structural diversity. Hum contact calls precede and grade into some of the food calls and occur as background "noise" from other vocalizing macaques

Lost Calls. Several lost calls were distinguishable, primarily according to differences in sound structure, amplitude, and manner of delivery.

Lost-1 moan: Typical of adult females and older juveniles, these calls could be heard for short distances beyond the boundaries of a group's normal area of dispersion. They had a moanlike quality, the higher pitched midsection of the call being more emphasized than the tails. Certain moanlike *hum* calls (Fig. 4A) were very similar to lost-1 moans (Fig. 4B), but differed in their much lower amplitude as well as in the context in which they were given.

Lost-2 yell: Compared to the low intensity lost-1 moans, it was shorter and louder and sounded like a yell rather than a moan (Fig. 4C).

Lost-3 yell: This loud call consisted of two parts or syllables: the first rose sharply in pitch and the accent was on the second syllable (Fig. 1D, E). There was virtually no

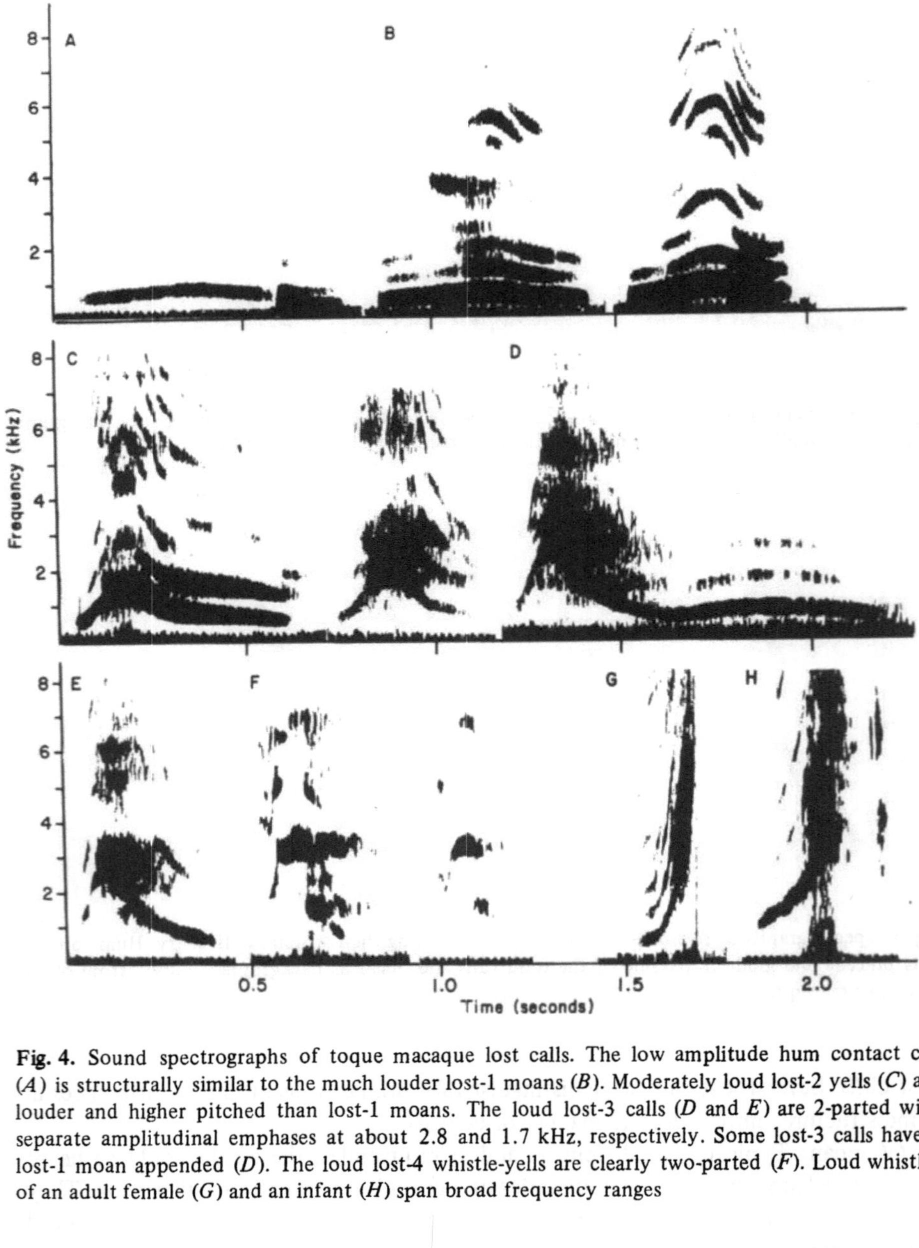

Fig. 4. Sound spectrographs of toque macaque lost calls. The low amplitude hum contact call (*A*) is structurally similar to the much louder lost-1 moans (*B*). Moderately loud lost-2 yells (*C*) are louder and higher pitched than lost-1 moans. The loud lost-3 calls (*D* and *E*) are 2-parted with separate amplitudinal emphases at about 2.8 and 1.7 kHz, respectively. Some lost-3 calls have a lost-1 moan appended (*D*). The loud lost-4 whistle-yells are clearly two-parted (*F*). Loud whistles of an adult female (*G*) and an infant (*H*) span broad frequency ranges

time gap between the end of the first and beginning of the second syllables; they were separated primarily by marked emphasis at different frequencies. Sometimes, a lost-1 moan was appended to this call (Fig. 4D). The call sounded higher-pitched than lost-2 yells.

Lost-4 whistle-yell: This two-syllable call was higher-pitched and more piercing than lost-3 yells. There was virtually no time gap between the two syllables, and their amplitudes were similar (Fig. 4F), but focused at different frequencies. The call was distinct, but, to the inexperienced human observer, it might be confused with the toque macaque's alarm call referring to large mammalian predators.

Lost whistle: This was a brief, high-pitched, piercing, almost pure-toned whistlelike probably less than 0.5 kHz, they were loud, approximately 0.20 s in duration, and atonal in quality. No spectrograph is available.

Lost whistle: This was a brief, high-pitched piercing, almost pure-toned whistlelike call. Whistles in infants and young juveniles (Fig. 4H) were higher pitched than those of adults (Fig. 4G).

Infantile Separation Calls. Sep-1 and *Sep-2* calls: There were at least two related kinds of these calls, distinguished mainly by the degree of modulation in frequency and amplitude. *Sep-1* calls were of relatively low amplitude and of more or less constant pitch (Fig. 5A). Onomatopoeically, they might be described as *coo* or *coo calls*. With increasing loudness a slight inflection (up to 1.0 kHz) developed in the middle of the call (Fig. 5B); this *sep-2* call had a slight moanlike quality. Infant *sep-2* calls were similar to the *lost-1* moans of adults, though of higher pitch, and probably were ontogenetically related to them.

Infantile grunts. These very quiet atonal vocalizations (Fig. 6) seemed similar to the grunts of older macaques (Fig. 2), but were of higher pitch.

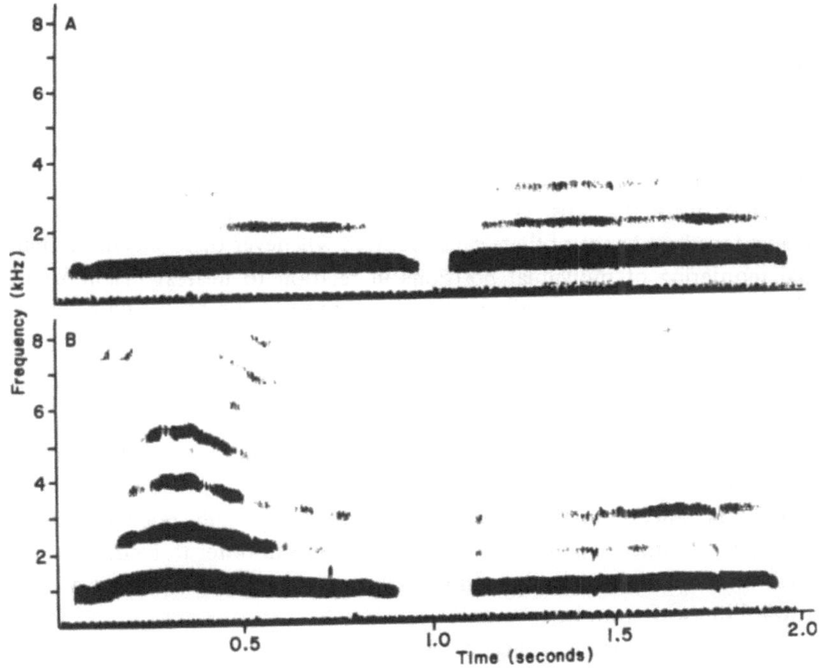

Fig. 5A,B. Sound spectrographs showing structural variations in the sep-1 or coo separation vocalization of infant toque macaques (**A**) and sep-2 moan separation calls (**B**)

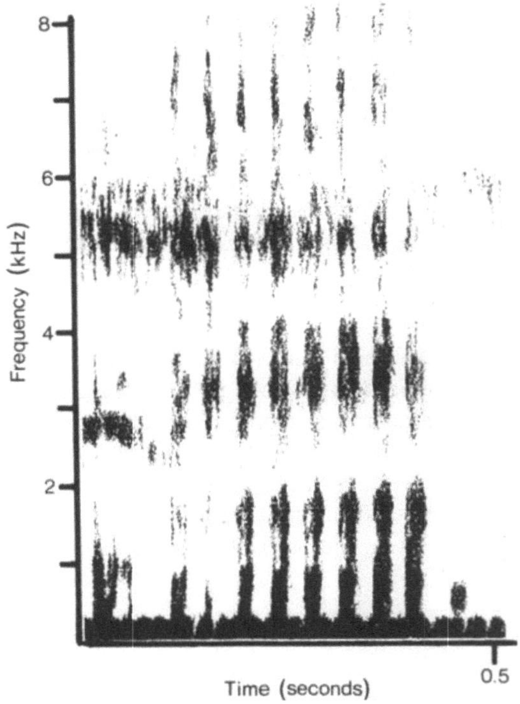

Fig. 6. Sound spectrographs of grunts of infants and young juveniles. Background noises at about 2.8 h and 5.2 kHz should be ignored

2.2 The Context and Significance of Call Types

Hum and Grunt Contact Calls. Hums and grunts occurred together in the same social contexts. The same individual often used both calls in sequence, although some individuals were predominantly hummers, whereas others were grunters.

Hums and grunts were the most commonly used vocalizations in the repertoire. Generally, a heightened exchange of these calls occurred whenever there was more than normal spatial reordering of individual positions in relation to nearest neighbors in the group. Thus, bouts of contact calls were always given in the following types of situations:

1. When animals initiated progression away from a resting place to set out on their new foraging route.
2. When an individual, or especially a subgroup, that had been separated from the main group, reunited with it.
3. When animals first encountered a new food source and spatially relocated themselves, in a hurried manner, among their nearest neighbors in order to feed.
4. At the onset or conclusion of an encounter with a neighboring group that involved running toward or away from it.
5. After group members stopped fleeing to rest quietly and to alertly scan their surroundings, or resumed normal activity following a predator alarm, which involved a rapid scattering of individuals for refuge.

Sometimes contact calls were also given in response to an auditory or visual disturbance external to the group, such as an alarm call from a neighboring group.

Once the group members settled into a set pattern of activity, such as a slow progression while foraging, contact calls were given only occasionally, and mostly antiphonally. The dialogue often increased (1–5 calls/min) among small parties of individuals who were split from the group, or who were foraging peripheral to it, and when visibility was poor.

An increased rate of exchange of slightly louder than normal hums and grunts usually indicated a local drawing together of animals (and increased feeding rate) at a moderately rich food source, such as a small fruited shrub. At very abundant food sources, such behavior led to food calls.

The most obvious response to most hums and grunts was either no marked response, or, more usually, a vocalization in kind, though a heightened exchange, as described above, usually attracted others to the call site.

Food Calls. Detailed quantitative analyses of food-calling behaviors have been published earlier (Dittus 1984). Food calls were given almost exclusively in the context of foraging (searching for, acquiring and ingesting food). Of 169 observations of food-calling events, 98% involved the discovery of an abundant food source. Typically, calls were given by one or more individuals as they first discovered a new source of superabundant food, such as a fully fruited tree. The call drew others to the site of the call. Calls ceased to be given within 5–10 min after discovering the source (after most of the group had been attracted), though feeding there often continued for 30 to 120 min.

Food calls were elicited by a variety of different food species and types, including fruits, flowers, herbs, mushrooms, rhizomes, and caterpillars, but they were uttered only at sites with a great abundance of these foods.

The most conspicuous response to food calls was to run to the site of the call to feed. The mere sound of the call, even at a great distance from the call site, was sufficient to elicit the response. Macaques also responded with one or more quiet hums and grunts as they approached the call site. While feeding at the source, a food call by one individual often triggered food calls and hums and grunts from others. The arrival of new group members at such sites, especially of dominant males, also triggered a round of these call types.

Infant Separation Calls. Four infant vocalizations are considered here: quiet *sep-1* or *coo* calls, *sep-2* calls, quiet *grunts*, and loud *whistles*.

Sep-1 calls were given by infants and young juveniles most often when visually isolated from any of their group members, and often when in visual contact with other group members, but after having been out of physical contact with their mothers for a prolonged period. Infants and young juveniles also called when an environmental obstruction, such as a wide gap in the tree canopy, prevented them from following or approaching group members. When prevented from nursing, infants also *coo* called quietly even though they were resting next to or in physical contact with their mothers. Especially during weaning, infant sep-1 or coo calls were interspersed with frustation "eich" vocalizations. The latter are not described here.

Infants that were truly separated from their group also gave sep-1 calls, but in addition, uttered the slightly louder sep-2 moans (Fig. 5B) and occasionally whistled (Fig. 4H). The latter call types were not given by infants seated close to others. The stimulus for uttering these calls appeared to be auditory isolation in addition to visual isolation. For example, tame captive infants whistled or moaned loudly only when they could not see or hear their human companions. Isolated infants also uttered very quiet grunts (Fig. 6), but these (or very similar ones) were also given by infants and young juveniles, for example, at the onset of group feeding sessions, when they were trying to feed close to others.

Older juveniles or adult macaques were never noted to direct sep-1 or sep-2 calls towards another macaque. However, very habituated macaques sometimes directed sep-1 calls at humans who were feeding them. Such calling accompanied typical begging behaviors (e.g., reaching towards the donor).

Infants uttering sep-1 calls, and who were in visual contact with, or otherwise fairly close to the mother and other group members, often elicited no vocalization or approach from others. But such infants were immediately retrieved by their mothers with the approach of a potential predator or another group of macaques. On the other hand, *sep-2 calls* and *whistles* from truly separated infants immediately triggered calling and searching from their mothers and often others as well.

Adult Lost-Calling Behavior. The social situations, in which *lost calls* were initiated, invariably and exclusively involved some degree of auditory and visual separation of one or more individuals from their group.

"Lost" individuals tended to rest at vantage points high in a tree from where they would look and call in a specific direction. Or they combined this behavior with active movement through the canopy, alertly scanning and calling in one or more directions, in a manner suggesting that they were searching for their group. When a small party was split from the group, one individual (usually an elder adult female) was the most active in calling and scanning the forest.

Mothers who had lost their infants, especially unweaned ones, frequently spent a good portion of the day lost-calling and searching, sometimes going on brief excursions of more than 100 m away from the group. Some mothers called and searched for several days before their behavior gradually returned to normal.

Within any one sequence of lost-calling behavior, the intensity of calls, and call types varied. The most frequent calls were moderately loud *lost-1* moans and *lost-2* yells and, to a lesser extent, *lost-3* calls that on some days were given at varying intervals over many hours. Quieter calling (*lost-1* moans) most often accompanied other activity, e.g., when a female gave up her more active search temporarily in order to feed. Or, lost-1 moans were simply given in between forays of more determined calling and searching. Loud *lost-4* calls were less often heard and not for more than a few minutes at a time. Bouts of *whistles* were the rarest and shortest (less than 1 min in duration). Along with other intense calls, whistles were given in response to the *lost* calls from the separated portion of the group.

In all cases, calling and searching behavior ceased with the reunion of the separated parties.

Nonvocal responses to *lost calls*: A frequent response of others, upon hearing the lost call of one of their separated adult group members, was initially a few quiet hum and grunt contact calls exchanged among group members, followed by silence. These quiet vocalizations were not audible by the lost member. However, such a virtually silent response often was accompanied by exploratory behavior from, usually, an adult male in the group. Typically, this behavior involved alertly looking in the direction of the caller, moving toward its calls through the forest canopy, climbing to an exposed vantage point, and simply sitting quietly or branch-shaking. Branch-shaking is primarily a long-distance visual signal but the sound of shaking branches can carry for short distances. Such essentially nonvocal responses by group members, by themselves, often provided sufficient cues for the lost caller to find and return to its group.

Lost adult individuals (usually adult males) occasionally called for long periods and the most evident response of fellow group members hearing the calls was to ignore the caller.

Vocal responses to *lost calls*: Frequently, the lost calls of separated individuals were responded to in kind. The initial reestablishment of vocal contact between separated parties usually involved a brief exchange (about 5–10 s) of intense calling and call types (lost-4, lost-5) including one or a few whistles. As the subgroups moved toward one another a similar vocal exchange was sometimes repeated. Upon reestablishing visual contact and reuniting, members of both parties again burst into an intense exchange of predominantly hum and grunt contact calls with a few low intensity *lost-1* moans. At such reunions, high-pitched squeals (similar to food calls) were also sometimes given.

2.3 Vocal Differences in Age, Sex, and Individual Identity

To the human ear, macaque vocalizations differed according to age and sex. Within each type of call, these differences are related mainly to pitch, adult male calls being of slightly lower pitch than calls of adult females, and juvenile calls being higher pitched than those of most adult females. For example, compare the vocalizations of young juveniles to those of adults for *hums* (Fig. 1A, B), *grunts* (Fig. 2A, B), *moans* and *whistles* (Figs. 4 and 5). Some of these differences apparently were size-related, as the calls of similar-sized juveniles were not distinguishable by sex. The voices of very old individuals often were raspy and wavery, unlike those of younger adults. Macaques also differed in their use of call types: juveniles and adult females hummed more than they grunted, whereas adult males grunted more.

I could distinguish a few well-known individuals by their unique *food* calls and *lost* calls. Also, since vocalizations do differ according to age-sex class, individual identity in many cases was conveyed by default: for example, that of the adult male in one-male groups. Most toque macaque groups have only a few representatives of a certain size class, so that discrimination of individuals within classes was narrowed.

Do the macaques distinguish individuals by voice alone? The following observations suggest that they do. Food calls by certain individuals drew only their kin and some dominant males, suggesting that the macaques recognized the vocalizing individual. This was also true of some contact calls that solicited strong responses, such as at moderately rich feeding sites. However, it is unknown whether most contact calls were individually unique and recognized as such. Infant vocalizations were most actively responded to (e.g., approach and retrieve) almost exclusively by their own mothers and occasionally by older siblings. Individuals responding, vocally or otherwise, to the lost calls of separated females or juveniles tended to belong to the same matrilineage. Lost calls from members of neighboring groups elicited no response.

3 Discussion

3.1 Comparisons with Other Primates

Call types of other species, which are equivalent to those considered here for *Macaca sinica*, are listed in Table 2. As in *M. sinica*, *hum types* of contact calls also have been noted as the most frequent in the repertoires of other species (Table 2). The infant *coo* calls of other macaque species were noted primarily from infants that had been separated from their mothers. The reason for the lack of reports of *food calls* in other species have been discussed (Dittus 1984). *Lost calls*, too have been described rarely, perhaps because most studies of primate vocal behavior occurred in captivity. Also, habitat differences in the degree of visual barriers to group cohesion, as well as predator pressure, may influence the prevalence of lost calling behavior between species and field sites.

Table 2. Call types of other primates homologous to the cohesion calls of *Macaca sinica*

Call type of *Macaca sinica*	Equivalent call type in other species		
	Species	Call-type label	Reference
Hum	*Macaca mulatta*	Clear call	Rowell and Hinde 1962
	M. fuscata	Type A1, A2	Itani 1963
		Class II coo call	Green 1975
		Coo call	Mitani 1986
	M. nemestrina	Coo call	Caldecott 1986
Grunt	*Cercopithecus aethiops*	Grunt	Struhsaker 1967
		Grunt	Cheney and Seyfarth 1982
	C. cephus	Grunt	Gautier 1978
Food call	*M. fuscata*	(Type A12)?	Itani 1963
	M. sylvanus	Food call	Personal observation
Infant	*M. mulatta*	Clear call	Rowell and Hinde 1962
Sep-1 and		Coo call	Lindburg 1971
Sep-2	*M. fuscata*	Coo call	Green 1975
Coo call	*M. arctoides*	Coo call	Lillehei and Snowdon 1978
	M. fascicularis	Flöten	Angst 1974
Lost call	*M. mulatta*	Cew, lost call	S.A. Altmann 1962
		Lost call	Lindburg 1971
	M. fuscata	Type A7, A8	Itani 1963
	M. arctoides	B calls	Bertrand 1969
	M. radiata	Lost call	Personal observation

3.2 The Significance of Individual (or Class) Recognition in Call Types

Several primates recognize their fellow group members individually by their voices. These include, among long-range calls, the *pant-hoot* calls of chimpanzees *(Pan troglodytes)* (Marler and Hobbett 1975), the intergroup *whoop-gobbles* of grey-cheeked mangabeys *(Cercocebus albigena)* (P.M. Waser 1977), and the *long calls* of cotton-top tamarins *(Saguinus oedipus)* (Snowdon et al. 1983; Snowdon and Hodun 1985); and among other calls, the *coo* calls of young stumptail *(Macaca arctoides)* and Japanese macaques *(M. fuscata)* (Lillehei and Snowdon 1978; Pereira 1986), two types of contact calls of pygmy marmosets *(Cebuella pygmaea)* (Snowdon and Cleveland 1980), isolation calls of infants squirrel monkeys *(Saimiri sciureus)* (Kaplan et al. 1978; Symmes et al. 1979; Symmes and Biden 1985), maternal voices of rhesus monkeys *(M. mulatta)* (Hansen 1976), and infant screams of vervet monkeys *(Cercopithecus aethiops)* (Cheney and Seyfarth 1980). In toque macaques *(M. sinica)*, too, females distinguish the agonistic screams of their own offspring from nonoffspring on the basis of the acoustic signal alone (Dittus and Baker, unpublished). Waser (1978) described the distinctive qualities of an old female mangababy's voice, and Gautier (1978) showed that several vocalizations of *Cercopithecus cephus* vary consistently between individuals, and many vocalizations change noticeably with age; but the discrimination of these differences by other monkeys was not tested. Similar individual structural variations have been found in the songs of several species of gibbons *(Hylobates)* (Haimoff and Tilson 1985; Haimoff and Gittins 1985; Tenaza 1985).

Two recent findings are of special interest to the present discussion. They concern the *cohesion* calls of ring-tailed lemurs *(Lemur catta)*, and especially the *coo* contact calls of Japanese macaques *(M. fuscata)*. These calls encode the caller's individuality and are used most often during foraging (Macedonia 1986; Mitani 1986). These calls of *M. fuscata* are homologous to the *coos* or *hums* of other members of the genus.

Green (1975) indicated that some of the *coo* contact calls, given by subordinate *M. fuscata* to dominant ones, differ structurally from the calls of dominant to subordinate macaques, although each sound type also overlapped other social contexts. In *C. aethiops*, Cheney and Seyfarth (1982) showed that responses to playbacks of recorded grunts toward subordinates differed from those of grunts toward dominant animals.

The data from these two studies suggest that different grunts or *coos* may function either as mild threats or as expressions of affiliative intent, although the original authors do not state this explicitly. The communicative value of these signals, nevertheless, would be dependent upon additional cues (vocal or visual) with regard to their source. For example, a mildly threatening grunt by an infant differs in social significance from that by an adult male. It seems most parsimonius to hypothesize that many *contact* calls identify the vocalizer as an individual or at least its age-sex class, as already suggested by Caldecott (1986) and Mitani (1986).

If cues concerning individual identity, or class, are imbedded in a vocal signal, then socially relevant information that is associated with animals as individuals (or types of individuals) is also accessible to listeners through their memories of past interactions. In short, it would inform a listener, if, in its unique history with the caller, the caller had been tolerant and protective, friendly and approachable, submissive and exploitable, aggressive, or whether it is strange and potentially dangerous. Individuals would be guided by such knowledge in selecting their social partners and their behaviors toward them.

3.3 The Social and Ecological Functions of Contact Calls

In the traditional view, *coo* contact calls function primarily to inform other group members of each others' relative locations and thus to contribute to their coordination and cohesion during progessions (Itani 1963; Lindburg 1971; Green 1975; W.I. Smith 1977; Macedonia 1986).

Such *contact* calls may have additional significance, however, when examined from an ecological perspective. Thus, I hypothesize that contact calls serve at least two functions: first, they guide individuals in selecting their social partners, especially during foraging; second, they may also constitute referential signals imparting information concerning the nature of food resources in the environment. Both qualities of *contact* calls would contribute to individual foraging success.

Communicating individual or class identity can be envisioned as being of a selective advantage in a variety of social and ecological contexts. However, it would be particularly important in the competitive and energetically expensive activity of food-finding. For example, *M. sinica* at Polonnaruwa and other primates spend 30 to 90% of daytime activity in the search for and consumption of food and water (Dittus 1977a; S.A. Altmann and J. Altmann 1970; Struhsaker 1975; Terborgh 1983; Robinson 1986). During the nonmating season, approximately 82% of all agonistic events relate to competition over food resources (Dittus 1977a). When foraging, individuals either exploit (or are exploited by) their nearest neighbors, by displacing them from feeding sites, or they tolerate each other feeding in proximity, and thereby share localized resources. Individuals choose their nearest neighbors in a manner to optimize their competitive ability in foraging. For example, individuals tend to avoid dominant animals who are feeding, and associate with those (kin, peers, or "friends") who tolerate them (Dittus 1977a, 1987; Baker-Dittus 1985; Caldecott 1986; Mitani 1986). Among primates that have been studied in their natural habitats, *hum* and *grunt* contact calls are most frequently heard at the onset of foraging progressions, when animals select their foraging partners, or at other times when they shift their relative positions during progressions, and when vision is obstructed (Itani 1963; Struhsaker 1967; Lindburg 1971; Caldecott 1986; Mitani 1986).

Contact calls would contribute directly to individual foraging success because they inform individuals of the identity and location of their nearest neighbors with greater efficiency than alternative behaviors would. For example, an individual can behave in two ways when it wishes to shift its position to feed at a new site among its neighbors, whose identity and/or location is uncertain. First, it can expend energy in approaching a neighbor, only to discover that it is a dominant individual near whom it is inhibited from feeding, and it may risk further energy loss and even injury if an attempt to feed there leads to a chase or fight. Alternatively, it can utter a contact call in order to solicit vocal responses from its neighbors, and thus it gains knowledge of the most suitable partners near which it can feed without risking energy loss or injury. Similarly, for dominant individuals, cuing in on the call to someone engaged in feeding, and going there to share or usurp the caller's food may be less costly than seeking a new feeding site (Dittus 1977a). Or the call of a dominant individual may cause others to avoid it, thereby enabling it more easily to monopolize a feeding site without wasting energy or risking injury in an agonistic bout.

The apparent function of contact grunts and hums as very mild threats or contact-promoting signals (cf. above) strengthens the hypothesis that such calls are important in fostering individual competitive ability in foraging.

3.4 Food and Contact Calls as Referential Signals Designating Differences in Food Resources

Food calls are referential signals advertising the presence and location of a large quantity of food in the environment. Dispersed group members, hearing food calls, run to feed at the rich sources announced by these calls (Dittus 1984). Hum and grunt contact calls always accompanied food-calling behaviors and food calls themselves appear to be structurally derived from hum contact calls.

Increased amplitude and rate of humming and grunting at a particular site often indicate a moderately abundant food source to which a few members were attracted to feed briefly. Food calls were absent at such sites; they were employed only if the food source was superabundant such that all or most group members could feed there for as long as 90 min per session, and for repeated sessions over 1 or more days (Dittus 1984).

Apparently these different vocal behaviors designate different degrees of food abundance in the environment. Future research might test whether subtle variations in humming, grunting, and food calling impart qualitative and/or quantitative differences concerning the nature of food resources, as has been found in the vocal behavior of domestic chickens (Marler et al. 1986). Such properties would be consistent with the proposed function of these calls in promoting foraging partnerships, competitive efficiency, and individual foraging success.

3.5 Dual Structure and Function of Adult Lost Calls

Lost calls advertise that a caller is separated from one or more of its close associates, and that it awaits from its lost associates a response communicating visual and/or auditory locational cues that would allow it to find and reunite with them.

Might lost calls have special physical properties that enhance their communicative efficiency over a long distance, as has been shown in other animal signals, for example, by Morton (1975); P.M. Waser and M.S. Waser (1977); and Brown and Waser (this Vol.)?

The physical structure of the toque macaque's lost calls and the manner in which they are used suggests that different lost calls serve slightly different, but complementary, functions depending upon the social context. First, most sound energy of the *lost-1* and *lost-2* calls (Fig. 4B, C) is concentrated between 500 and 1800 Hz, and they have relatively long durations (0.30–0.75 s). These calls are often given for many minutes or hours at varying rates of delivery. They appear to be physically designed for long-distance transmission with relatively few directional cues. They seem to function as sound beacons, more or less chronically broadcasting the "state of being lost" and seeking a response from the separated party. As subgroups move independently of each other through their home range, such broadcasting would increase the chances of establishing auditory contact between them.

Second, the main energy emissions of the *lost-4* call and whistle are near 3500 Hz or past 5000 Hz, respectively, and are therefore considerably higher than the optimum of 500–2000 Hz for distant propagation. Also, these calls are shorter in duration than the lost-1 and lost-2 calls (Table 1). However, both calls seem to be especially suited to facilitate their *directional* location by a distant receiver. The abrupt boundaries of both syllables in the lost-4 call, and the accentuation of this abruptness through their immediate temporal juxtaposition and marked frequency drop of the second syllable makes this call suitable for comparisons in the difference in time of arrival of the sounds at the two ears of the receiver. The *whistle*, too, has sharp beginnings and ends, and spans a broad frequency range. Both call types are often repeated several times in succession and may be especially adapted to communicate directionality over a long distance (Marler 1955, 1959). These calls are most commonly uttered, antiphonally and briefly, upon first reestablishing vocal contact between distantly separated parties, i.e., at a juncture where directional cues would be most appropriate for orientation to guide a reunion. *Lost-3* calls appear to combine properties of the lost-2 and lost-4 calls. The loud atonal pulsed properties of the lost-5 calls would facilitate localization at a distance.

This interpretation of the dual function of toque macaque lost calls agrees with studies of pygmy marmosets *(Cebuella pygmaea)* and squirrel monkeys *(Saimiri sciureus)*. In these species it had been demonstrated that *contact* or *isolation* calls used at far distances possess more physical cues for sound localization than those used at close distances (Snowdon and Hodun 1981; Masataka and Symmes 1986).

3.6 Triple Structure and Function in Infantile Separation and Lost Calls

The *grunts* of infants are barely audible by humans further than 20 m from their source. Their low-pitched, rapid trill quality, however, makes their source easy to locate within short distances. Among infants the *sep-1* and *sep-2* calls were the least locatable because they fade in and out, lacking sharp acoustic beginnings and ends, are fairly uniform in amplitude and frequency, and are pure toned (e.g., Marler 1955, 1959). These calls seem to have been modified in their evolution to thwart detection by predators; infants uttering these calls also remained in good cover fairly high above ground.

Infant separation calls appear to have slightly different, but complementary functions. The *sep-2* calls advertise that an infant is potentially split from the group, the sound carrying for a moderate distance (approximately equal to the normal dispersion distance of a group), but betraying a minimum of locational cues. A mother searching for her infant, however, would be especially attentive to her infant's vocalization, and might use its very soft *grunts* to find its exact location. The loud *whistles* impart the degree of severity of the infant's isolation and simultaneously give long-distance directional cues to guide its rescuers. Lost infants whistle most commonly upon hearing the vocalizations of other macaques, and their *whistles* always elicited an approach or retrieval.

3.7 Lost Calls in Relation to Intergroup Spacing Behavior

Vocalizations in other Old World monkeys that carry over long distances have generally been linked to territorial behavior, particularly among colobine monkeys and African forest monkeys (e.g., Ripley 1967; Marler 1972, 1973; P.M. Waser 1975). Hence, do lost calls in macaques function in intergroup spacing? Lost calls in *M. sinica* were not obviously responded to by members of other groups, nor were they associated with behaviors that typically involved intergroup avoidance or spacing behaviors, such as face-offs, peaceful displacements, fights, and chases between groups. The most common intergroup spacing behavior concerned branch shaking. No vocalizations accompanied branch-shaking; it was primarily a long-distance visual display, although the sound of cracking or falling dead branches did carry for some distance, and these sounds did sometimes elicit branch-shaking from neighboring groups. These observations, and the lack of vocalizations specific for intergroup spacing, are supported by reports of the behaviors of other macaques that had been studied in their natural habitats (Simonds 1965; Southwick et al. 1965; Lindburg 1971; Caldecott 1986).

4 Summary and Conclusions

1. The toque macaque's *(Macaca sinica)* contact calls, hums and grunts, are varied in physical structure, the most common vocalizations in the repertoire, and are used in a variety of social situations. They differ acoustically and structurally according to age and sex, and probably impart information concerning the vocalizer's location, age-sex class, and its individual identity. They are homologous to the individualistic "contact calls" or coo calls of Japanese and other macaque species.
2. *Food calls* designate a superabundant food source and attract most or all group members to feed at the call site. Food calls are structurally derived from hums and are always used in combination with hums and grunts. Some calling bouts with a high rate of humming and grunting, but lacking food calls, may referentially designate the location of a moderately abundant food source, to which only some group members go to feed.
3. I hypothesize that, individually unique hum and grunt contact calls guide animals in their more or less continuous selection of foraging partners. The identity of nearest neighbors during foraging influences the degree of interference (from dominant neighbors) or of food-sharing that an individual is subject to, and therefore influences its feeding efficiency. Contact calls are less costly than other behaviors potentially might be, in guiding individuals in their selection of foraging partners. Hence, contact calls would contribute directly to an individual's competitive efficiency and foraging success.
4. There are essentially two kinds of lost calls that complement one another in structure and function. The first are used more or less chronically to broadcast the "state of being lost" and a desire for a response from the separated party. The low frequency, pure tone, and long duration of these calls appear to be optimally adapted for propagation over long distances but with a minimum of directional cues. They have a moan-like quality and grade into loud yells.
5. The second concerns more or less discrete signal types that possess sharp temporal boundaries, and a broad frequency range with most energy being concentrated in the

upper frequencies. These properties, their two-syllabic structure, or their mode of delivery of several in succession would impart directional cues to a distant listener. To the human ear they are high-pitched *whistle-yells* and *whistles*. They are given predominantly antiphonally, and briefly at the juncture when distantly separated parties first reestablish vocal contact. Their directional cues seem to orient and guide the reunion of separated parties.

6. Among infants, low-pitched *coo* calls of long duration seem to advertise that an infant is potentially lost over distances approximately equal to a group's normal dispersion. Their indistinct beginnings and ends betray a minimum of directional cues, and infants giving these calls tend to hide in dense foliage. Directional cues are given by barely audible atonal *grunts* that may orient a searching mother at fairly close range. When an infant is distantly separated from its group, its loud *whistles*, mostly in response to the lost calls of its mother, give directional cues over long distances.

Acknowledgments. I thank the Office of the President of the Democratic Socialist Republic of Sri Lanka for permission to do basic research in Sri Lanka, and especially Prof. Hilary Crusz of the University of Peradeniya for his friendship and many years of selfless assistance in official matters. I am also very grateful for the hospitable and encouraging support from Prof. Cyril Ponnamperuma, Director of the Institute of Fundamental Studies, which sponsors the research under its aegis. Similar thanks go to Prof. N.C. Wickramasinghe. I thank Dr. Anne Baker, who assisted with fieldwork, Shirani Nathanael, who helped with data summarization, Prof. Dietmar Todt, who helped prepare the sound spectrographs, and Françoise Bayart and Shyamala Ratnayeke, who made helpful editorial comments. The research was financed at various stages by National Science Foundation grants, BNS-7619740, BNS-7924057, BNS-8300092, and BNS-8609665. Additional funding was provided by the Harry Frank Guggenheim Foundation, the Wenner-Gren Foundation for Anthropological Research, the National Geographic Society, and the Friends of the National Zoo. I thank Devra Kleiman for administrative assistance.

References

Altmann SA (1962) Social behavior of anthropoid primates: analysis of recent concepts. In: Bliss E (ed) Roots of behavior. Harper, New York, pp 277–285

Altmann SA, Altmann J (1970) Baboon ecology. Univ Chicago Press, Chicago

Angst W (1974) Das Ausdrucksverhalten des Javaneraffen *Macaca fascicularis* Raffles 1821. Adv Ethol 15:1–91

Baker-Dittus A (1985) Infant- and juvenile-directed care behaviors by adult toque macaques *(Macaca sinica)*. PhD Diss, College Park, Univ Maryland

Bertrand M (1969) The behavioral repertoire of the stumptail macaque. Bibl Primatol 11:1–273

Caldecott JO (1986) An ecological and behavioral study of the pig-tailed macaque. Contrib Primatol 21:1–259

Cheney DL, Seyfarth RM (1980) Vocal recognition in free-ranging vervet monkeys. Anim Behav 28:362–267

Cheney DL, Seyfarth RM (1982) How vervet monkeys perceive their grunts: field playback experiments. Anim Behav 30:739–751

Dittus WPJ (1977a) The social regulation of population density and age-sex distribution in the toque monkey. Behaviour 63:281–322

Dittus WPJ (1977b) The ecology of a semi-evergreen forest community in Sri Lanka. Biotropica 9:268–286

Dittus WPJ (1984) Toque macaque food calls: semantic communication concerning food distribution in the environment. Anim Behav 32:470–477

Dittus WPJ (1985) The influence of cyclones on the dry evergreen forest of Sri Lanka. Biotropica 17:1–14

Dittus WPJ (1986) Sex differences in fitness following a group take-over among toque macaques: testing models of social evolution. Behav Ecol Sociobiol 19:257–266

Dittus WPJ (1987) Group fusion among wild toque macaques: an extreme case of inter-group resource competition. Behaviour 100:247–291

Dittus WPJ, Thorington RW Jr (1981) Techniques for aging and sexing primates. In: Subcommittee on Conservation of Natural Primate Populations (eds) Techniques for the study of population ecology. Nat Acad Press, Washington DC, pp 81–131

Gautier J-P (1978) Repertoire sonore de *Cercopithecus cephus*. Z Tierpsychol 46:113–169

Green S (1975) Variation of vocal pattern with social situation in the Japanese monkey *(Macaca fuscata)*: a field study. In: Rosenblum LA (ed) Primate behavior. Academic Press, London New York, pp 1–102

Haimoff EH, Gittins P (1985) Individuality in the songs of wild agile gibbons *(Hylobates agilis)* of penninsular Malaysia. Am J Primatol 8:239–247

Haimoff EH, Tilson RL (1985) Individuality in the female songs of wild Kloss' gibbons *(Hylobates klossii)* on Siberut Island, Indonesia. Folia Primatol 44:129–137

Hansen EW (1976) Selective responding by recently separated juvenile rhesus monkeys to the calls of their mothers. Dev Psychobiol 9:83–88

Itani J (1963) Vocal communication of the wild Japanese monkey. Primates 4:11–66

Kaplan JN, Winship-Ball A, Sim L (1978) Maternal discrimination of infant vocalizations in squirrel monkeys. Primates 19:187–193

Lillehei R, Snowdon CT (1978) Individual and populational differences in the vocalizations of young stumptail macaques *(Macaca arctoides)*. Behaviour 65:270–281

Lindburg DG (1971) The rhesus monkey in North India: an ecological and behavioral study. In: Rosenblum LA (ed) Primate behavior. Academic Press, London New York, pp 1–106

Macedonia JM (1986) Individuality in a contact call of the ringtailed lemur *(Lemur catta)*. Am J Primatol 11:163–179

Marler P (1955) Characteristics of some animal calls. Nature (London) 176:6–8

Marler P (1959) Developments in the study of animal communication. In: Bell PR (ed) Darwin's biological work. Cambridge Univ Press, Cambridge, pp 150–206

Marler P (1972) Vocalizations of East African monkeys II: black and white colobus. Behaviour 42: 175–197

Marler P (1973) A comparison of vocalizations of red-tailed monkeys and blue monkeys, *Cercopithecus ascanius* and *C. mitis*, in Uganda. Z Tierpsychol 33:223–247

Marler P, Hobbett L (1975) Individuality in a long-range vocalization of wild chimpanzees. Z Tierpsychol 38:97–109

Marler P, Dufty A, Pickert R (1986) Vocal communication in the domestic chicken: I. Does a sender communicate information about the quality of a food referent to a receiver? Anim Behav 34: 188–193

Masataka N, Symmes D (1986) Effect of separation distance on isolation call structure in squirrel monkeys *(Saimiri sciureus)*. Am J Primatol 10:271–278

Mitani M (1986) Voiceprint identification and its application to sociological studies of wild Japanese monkeys *(Macaca fuscata yakui)*. Primates 27:397–412

Morton E (1975) Ecological sources of selection on avian sounds. Am Nat 108:17–34

Pereira ME (1986) Maternal recognition of juvenile offspring coo vocalizations in Japanese macaques. Anim Behav 34:935–937

Ripley S (1967) Intergroup encounters among Ceylon gray langurs *(Presbytis entellus)*. In: Altmann SA (ed) Social communication among primates. Univ Chicago Press, Chicago, pp 237–253

Robinson JG (1986) Seasonal variation in use of time and space by the wedge-capped capuchin monkey, *Cebus olivaceus*: implications for foraging theory. Smithsonian Contrib Zool 431: 1–60

Rowell TE, Hinde RA (1962) Vocal communication by the rhesus monkey *(Macaca mulatta)*. Proc Zool Soc London 138:279–294

Simonds PE (1965) The bonnet macaque in South India. In: DeVore I (ed) Primate behavior. Holt, Rinehart and Winston, New York, pp 175–196

Smith HJ, Newman JD, Hoffman HJ, Fetterly K (1982) Statistical discrimination among vocalizations of individual squirrel monkeys *(Saimiri sciureus)*. Folia Primatol 37:267–279

Smith WJ (1977) The behavior of communicating. Harvard Univ Press, Cambridge

Snowdon CT, Cleveland J (1980) Individual recognition of contact calls by pygmy marmosets. Anim Behav 28:717–727

Snowdon CT, Hodun A (1981) Acoustic adaptations in pygmy marmoset contact calls: locational cues vary with distances between conspecifics. Behav Ecol Sociobiol 9:295–300

Snowdon CT, Hodun A (1985) Troop-specific responses to long calls of isolated tamarins *(Saguinus mystax)*. Am J Primatol 8:205–213

Snowdon CT, Cleveland J, French JA (1983) Responses to context- and individual-specific cues in cotton-top tamarin long calls. Anim Behav 31:92–101

Southwick CH, Beg MA, Siddiqi MR (1965) Rhesus monkeys in North India. In: DeVore I (ed) Primate behavior. Holt, Rinehart and Winston, New York, pp 111–159

Struhsaker TT (1967) Auditory communication among vervet monkeys *(Cercopithecus aethiops)*. In: Altmann SA (ed) Social communication among primates. Univ Chicago Press, Chicago, pp 281–324

Struhsaker TT (1969) Correlates of ecology and social organization among African cercopithecines. Folia Primatol 11:80–118

Struhsaker TT (1975) The red colobus monkey. Univ Chicago Press, Chicago

Symmes D, Biden M (1985) Maternal recognition of individual infant squirrel monkeys from isolation call playbacks. Am J Primatol 9:39–46

Symmes D, Newman JD, Riggs G, Lieblich AK (1979) Individuality and stability of isolation peeps in squirrel monkeys. Anim Behav 27:1142–1152

Tenaza RR (1985) Songs of hybrid gibbons (*Hylobates lar* x *H. muelleri*). Am J Primatol 8:249–253

Terborgh J (1983) Five new world primates. Princeton Univ Press, Princeton

Vencl F (1977) A case of convergence in vocal signals between marmosets and birds. Am Nat 111: 777–782

Waser PM (1975) Experimental playbacks show vocal mediation of intergroup avoidance in a forest monkey. Nature (London) 255:56–58

Waser PM (1977) Individual recognition, intragroup cohesion and intergroup spacing: evidence from sound playback to forest monkeys. Behaviour 60:28–74

Waser PM (1978) Postreproductive survival and behavior in a free-ranging female mangabey. Folia Primatol 29:142–160

Waser PM, Waser MS (1977) Experimental studies of primate vocalization: specializations for long-distance propagation. Z Tierpsychol 43:239–263

Environmental Influences on the Structure of Primate Vocalizations

CHARLES H. BROWN[1] and PETER M. WASER[2]

Under natural conditions vocal signals are embedded in background noises, and are degraded by the action of diffraction, reflection, frequency-dependent absorption, and refraction. A fuller appreciation of the evolution of communication systems may be achieved by developing an understanding of the impediments to communication which occur in the natural environment, and investigating possible adaptations in the structure of either productive and receptive systems which may act to counter these impediments. The idea that characteristics of vocal signals have evolved a form acting to counter ecological impediments to communication has received increasing attention over the past 2 decades. Although most investigators have addressed questions pertinent to the structure of bird song (Morton 1970, 1975; Chappuis 1971; Jilka and Leisler 1974; Nottebohm 1975; Marten and Marler 1977; Marten et al. 1977; Bowman 1979; Hunter and Krebs 1979; Wasserman 1979; Gish and Morton 1981; Brenowitz 1982; Heuwinkel 1982; Wiley and Richards 1982; Shy 1983; Cosen and Falls 1984; McGregor and Falls 1984; McGregor and Krebs 1984; Ryan and Brenowitz 1985), other investigators have focused on the vocalizations of diverse groups of vertebrates including bats (Griffin 1971), frogs (Ryan 1986), whales (Payne and Webb 1971; Payne and Guinee 1983), ground squirrels (Brown and Schwagmeyer 1984), and nonhuman primates (P.M. Waser and M.S. Waser 1977; P.M. Waser and Brown 1984, 1986).

Selection may shape not only signal form (where more reliable or longer-distance transmission of information is advantageous), but also perceptual capabilities (where more complete or accurate reception of information is beneficial). In addition, selection may shape other aspects of communicative behavior, such as the timing of vocalizations or the choice of broadcast sites. In this chapter we briefly review some examples of possible environmental effects on perceptual capabilities and calling behavior; we then summarize the results of our studies on the acoustic ecology of equatorial African primate habitats, and finally describe in more detail our ongoing work focused on how habitat acoustics influence the fidelity of calls broadcast in natural environments.

1 Habitat Effects on Perceptual Abilities and Calling Behavior

Brown and P.M. Waser (1984) found that arboreal, forest-living blue monkeys are up to 20 dB more sensitive to low-frequency sounds (signals 500 Hz and lower) than are semiterrestrial, more open-country rhesus monkeys. This difference in sensitivity paral-

1 University of South Alabama, Mobile, Alabama 36688, USA
2 Purdue University, West Lafayette, Indiana 47907, USA

Todt/Goedeking/Symmes (Eds.)
Primate Vocal Communication

lels and is likely a consequence of differences in habitats where arboreal forest species may capitalize on a highly efficient low-frequency communication window which is unavailable to organisms resident in open habitats. If species differences can be found in perceptual capabilities as fundamental as audiograms, it seems likely that more sophisticated aspects of audio processing will also vary in ways that reflect habitat differences. In this respect it has already been found that rain forest monkeys [blue monkeys *(Cercopithecus mitis)* and grey-cheeked mangabeys *(Cercocebus albugena)*] are significantly better at hearing their own species-specific calls in simulated forest noise than they are at detecting the calls of other species (Brown 1986). Furthermore, forest monkeys hear monkey calls in noise at much lower levels than humans are able to hear speech (Brown 1986). These results suggest that forest-dwelling primates have evolved specializations in both their sound processing and sound production apparatus which aid communication in the acoustically noisy and visually restricted rain forest habitat.

Environmental factors have also been shown to reflect aspects of calling behavior. In many forest primates, the probability of emitting long-distance calls peaks sharply at or shortly after dawn; the peak is earlier or shorter in some species (e.g., black and white colobus species) than others (e.g., grey-cheeked mangabeys, and many forest *Cercopithecus* species: Gautier and Gautier-Hion 1977). Call transmission has been shown to be more efficient at this time of day; probably due to strong refraction from inversionlike thermal gradients (P.M. Waser and M.S. Waser 1977; Henwood and Fabrick 1979). In some semiterrestrial primates (e.g., baboons and agile mangabeys) long-distance calls are given primarily from elevated sites. Elevated calling also markedly increases transmission distance in some habitats (Morton 1975; Marten et al. 1977; Brown and Schwagmeyer 1984; P.M. Waser and Brown 1986).

Other aspects of signaling behavior may have also evolved under pressure from habitat acoustics. Some calls are highly chorused between groups (e.g., black and white colobus *roars*) or even species (e.g., blue monkey *pyows* or red-tail monkey *hacks*: Marler 1973; Gautier and Gautier-Hion 1977). Does contagious calling increase the audible range of signals, or the likelihood that recipients will be listening? Many imaginable environmental effects on primate calling behavior remain quite uninvestigated. Do some primates, like some birds and frogs, issue their calls during the silent intervals between the vocal bouts of conspecifics? Are calls given more frequently or more repetitively in noisy environments, as might be expected given the relationship between redundancy and the reliability of communication (Potash 1972)? Are "spontaneous" calls given when vocalizers have reached particularly good calling sites, or simply new calling sites from which they are more likely to influence new recipients?

2 Habitat Acoustics

It is likely that differences in the structure of the vocalizations of various species are due to differences in the acoustic ecology of their respective habitats. This possibility is predicated on the assumption that different primate habitats differ acoustically. P.M. Waser and Brown (1986) explored this possibility by systematically conducting measurements of the acoustic ecology of three classes of East African primate habitats: rain forest, riverine forest, and savanna.

2.1 Background Noise

Habitat differences may have influenced both the modality of choice and the structural complexity of communication signals (Marler 1965, 1967, 1968). Because vocalizations must be detected in ongoing and fluctuating background noise, one key feature of the acoustic environment is the level and spectrum of the background noise. P.M. Waser and Brown (1986) measured the ambient noise of representative East African primate habitats as a function of time of day. Figure 1 shows the ambient noise level of each of the three habitat types measured in third-octave bands from 06.00 h to 18.00 h. Overall noise levels were lowest in savanna and highest in riverine forest. Because these differences were both consistent and significant, ambient noise characteristics may have influenced the structure of vocal signals given by species resident in each class of habitat. In rain forest, noise levels varied prominently with frequency at all times of day; levels were consistently high below 100 Hz, and between 2000 and 4000 Hz. Rain forest noise in the 2000 to 4000 Hz region, averaged across time of day, exceeded that in riverine and savanna habitats, while noise in the 200 Hz to 1000 Hz region was as quiet or quieter in rain forest than in the comparison habitats.

The origins of noise differed between habitats as did the resulting noise spectra. In the savanna biotic sources do not contribute significantly to ambient noise levels except at 06.00 h where high-frequency insect noise (4000 Hz) was evident before sunrise and the subsequent elevation of wind levels. Exhaustive measurements of wind patterns in Amboseli Park, Kenya (J. Stelzner, personal communication) have shown that solar heating of the plains results in an average wind velocity increasing from 0.95 m s^{-1} at 09.00 h to 2.61 m s^{-1} at 17.00 h. In rain forest, the diurnal fluctuation in noise within the 2000 to 4000 Hz band was principally due to the activities of birds and insects, while low-frequency noise was probably due to vegetation movement and dripping condensation. Noise in riverine forest includes significant components from both biotic sources and from the interaction of forest vegetation with wind sweeping off the savanna. These acoustic differences documented between habitats are substantially greater than differences within habitats (i.e., differences between two rain forest sites, for example, have been shown to be small: Brown and P.M. Waser 1984). Ambient noise characteristics are then one feature of the acoustic ecology which differ between habitats and which could act as a source of selection resulting in divergence in the structure of vocalizations given by organisms resident in different habitats.

Though riverine forest tests revealed the highest measured noise levels, P.M. Waser and Brown (1986) reported that their greatest subjective impression of noise was experienced in the savanna, and that this noise was produced by the interaction of wind with their pinnea and auditory systems. Receptive wind noise has been shown by Kristiansen and Pettersen (1978) to markedly impair hearing in human listeners tested in a wind tunnel. Turbulence in airflow around the head produces low-frequency noise that peaks at higher levels and at higher frequencies for higher wind velocities (e.g., 90 dB and 200 Hz at 0.5 m s^{-1}). Whether adaptations of the shape of the head and external ear act to minimize hearing difficulties in open-country mammals remains unknown.

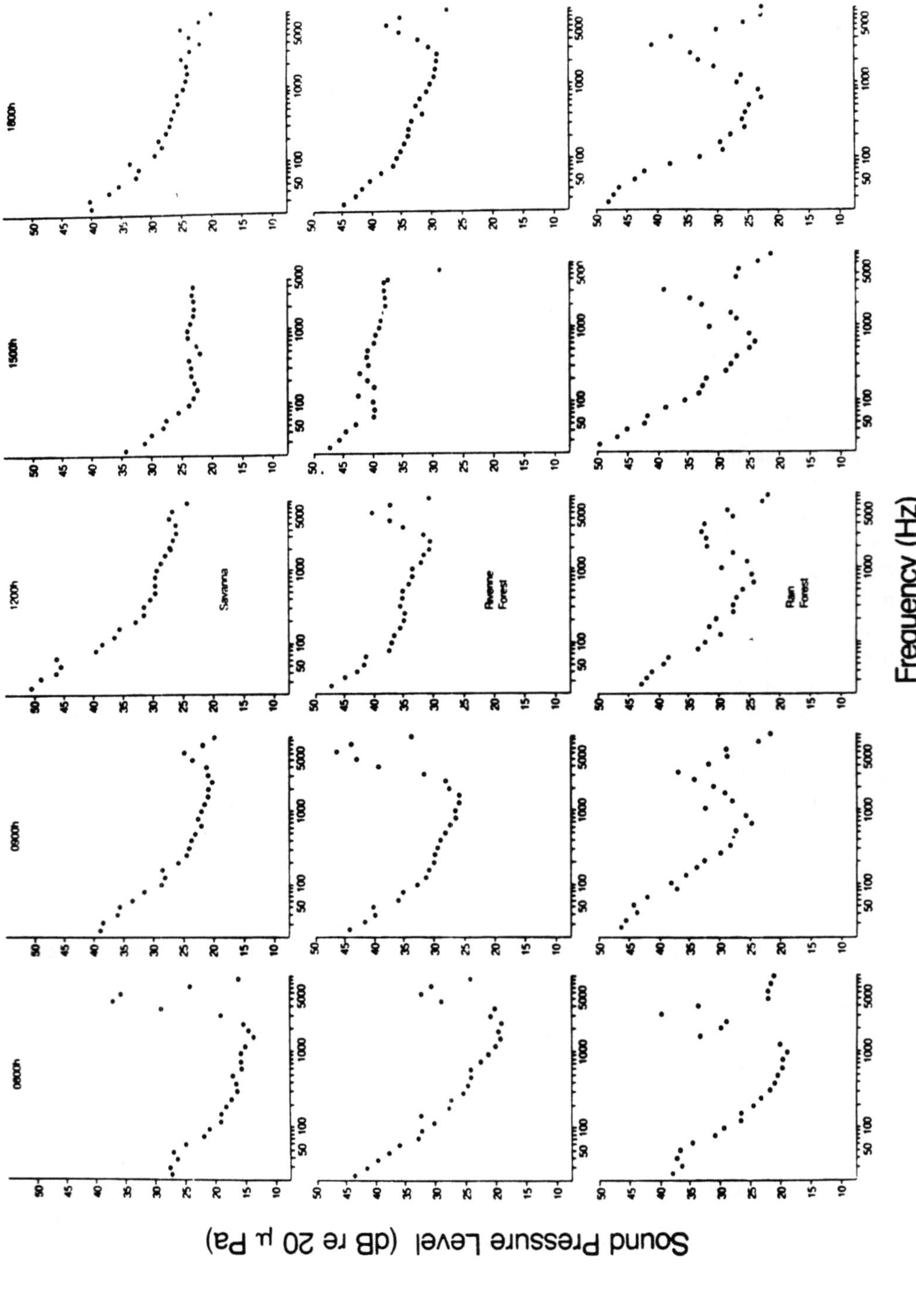

Fig. 1. Ambient noise according to time of day recorded in third-octave bands (data from P.M. Waser and Brown 1986)

2.2 Sound Propagation

Sound propagation tests conducted with appropriately calibrated signals (Brown and P.M. Waser 1984; P.M. Waser and Brown 1984, 1986) show that sound propagation differs between habitats and between test frequencies within a habitat (Fig. 2). Sound propagation is measured as excess attenuation. Because sound commonly radiates spherically in all directions from its source, the amplitude of the wave front obeys the inverse square law. Excess attenuation is the extra decrement in signal amplitude that exceeds that expected from the inverse square law. This excess in attenuation can be due to atmospheric absorption or to diffraction and absorption of the sound by environmental surfaces. Over the typical distances at which primate utterances are audible atmospheric absorption is a minor contributor to excess attenuation.

Excess attenuation was more marked in savanna and riverine forest tests than in the rain forest. The differences were particularly evident in the frequency range between 200 and 2000 Hz. Near the ground in both savanna and riverine forests, excess attenuation tended to increase with an increase in test frequency, and no band of frequencies was propagated very favorably. Furthermore, excess attenuation values in these habitats were more variable over frequency than those in rain forest. In rain forest, at 200 Hz, the attenuation rate was less than that given by the inverse square law. Hence, a "sound window", or a preferred frequency band for superior long-range propagation was observed. This outcome was likely due to an inversionlike thermal gradient acting to alter the ray characteristics of the direction of propagation of the wave front such that rays directed toward the upper canopy are bent back toward the lower strata which are occupied by primates (Brown 1986). Appropriate gradients have been measured across tropical forest canopies, particularly in the first few hours after dawn (P.M. Waser and M.S. Waser 1977). The channeling of sound in the rain forest is reminiscent of that which occurs in the ocean in the sofar zone (Webb and Tucker 1970). In both cases long-range acoustic communication signals may have evolved to capitalize upon this characteristic of the environment. Furthermore, the adaptation is likely to be behavioral as well as structural, because species vocalize in the appropriate strata (canopy, time of day, or sofar zone), as well at the appropriate frequency.

In all three habitats, the increase in excess attenuation with increasing propagation distances was approximately logarithmic; that is, excess attenuation increased approximately a constant amount per doubling of distance. The logarithmic relationship between excess attenuation and propagation distance is displayed in Fig. 3.

The three classes of primate habitats differ both in background noise and in sound propagation characteristics; these differences have consequences for vocal communication over both short and long distances. Over short distances, signals propagate equally well in all habitats (Fig. 3). Near the source, signal-to-noise ratios are influenced more by habitat differences in background noise than by habitat differences in propagation rates. As a result, it is likely that the form of short-range signals should more closely reflect a habitat's background noise characteristics than its attenuation properties. Furthermore, relatively loud short-range signals (those with a large signal-to-noise ratio) should be free to exhibit a form which is independent of the constraints imposed by habitat acoustics. Conversely, signals designed to function at the limits of audibility should exhibit a form reflecting habitat-specific constraints.

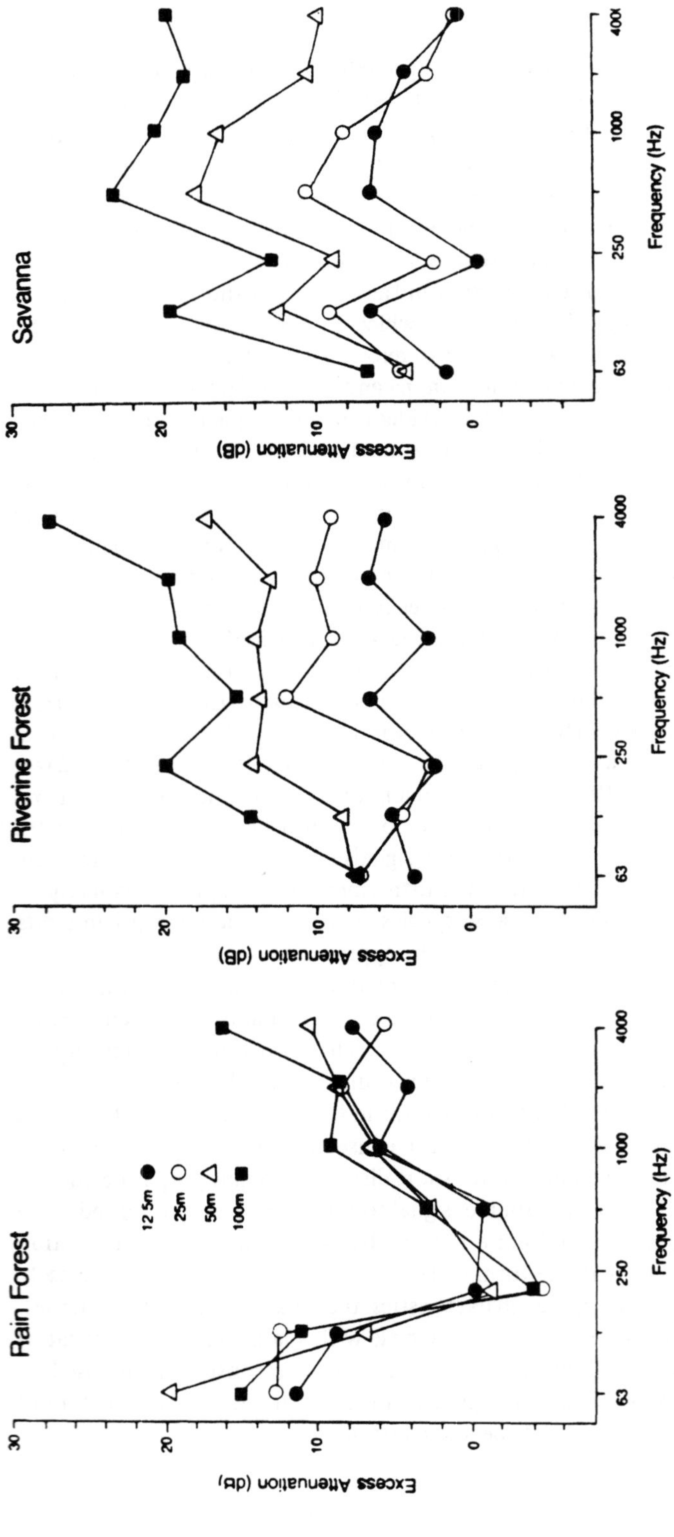

Fig. 2. Excess attenuation of broadcast tones as a function of source-receiver distances of 12.5, 25, 50, and 100 m (data from P.M. Waser and Brown 1986)

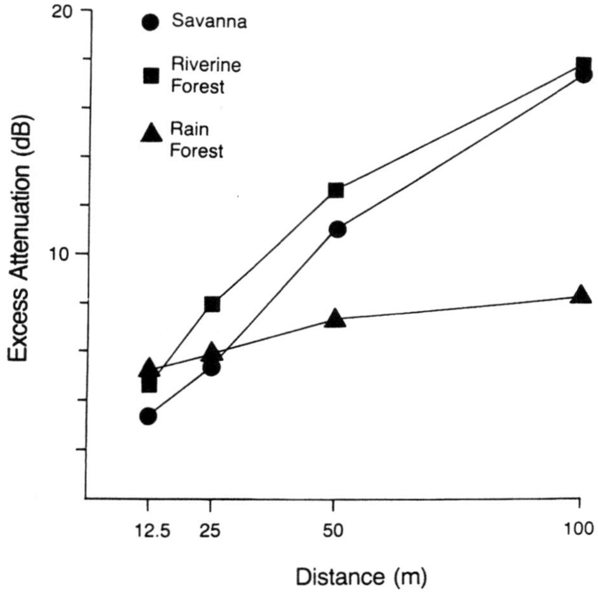

Fig. 3. Excess attenuation of tones, averaged across frequency in each habitat, as a function of broadcast distance (data from P.M. Waser and Brown 1986)

At distances of 50-100 m, habitat-specific differences in sound transmission become prominent, and equal or outweigh differences in background noise (Fig. 3). At these distances, the details of excess attenuation vary complexly with frequency and distance (Fig. 2), and this complexity may make it possible for natural selection to have favored calls with certain features for transmission over specific distances. In riverine forest, for example, signals 200 Hz in frequency propagate well to 25 m, but are not propagated well beyond that. This complexity in attenuation characteristics makes it difficult to specify what call structures would be favored for communication over these intermediate distances. One way to circumvent this problem is to broadcast calls and call variants in "appropriate" and "inappropriate" habitats and to conduct measurements of the growth of distortion of the call with transmission distance in each habitat. Our current measurements of relative call distortion are described in a following section.

At transmission distances of 100 m or more, habitat differences in attenuation are prominent, and are great enough to outweigh habitat differences in background noise (P.M. Waser and Brown 1986). Hence, the form of long-range signals should more closely reflect the attenuation characteristics of the habitat than its background noise characteristics.

2.3 Other Impediments to Vocal Communication

Background noise and attenuation are but two factors constraining acoustic communication. A call may remain audible after information coded by temporal or frequency modulation has been obscured by environmental variables. Wiley and Richards (1978, 1982; also see Richards and Wiley 1980) have focused on two critical signal-degrading processes: amplitude fluctuations and reverberation. Amplitude fluctuations are apt to occur when sound is scattered from moving objects, such as turbulent air, and would be expected to be more prominent in open habitats. However, empirical measurements of amplitude fluctuations of broadcast tones indicate that fluctuations in the rain forest are potentially as troublesome as those in savanna and riverine habitats (P.M. Waser and Brown 1986). In all cases, these fluctuations have the capacity to seriously degrade the transmission of information coded by amplitude modulation, especially at higher carrier frequencies and transmission distances. Under these conditions, like a distant radio station, the information bearing feature of a signal may fade in and out of reception.

Reverberation is the decay of abrupt signal onsets and offsets due to the scattering and reflection of sound from multiple surfaces in the habitat. Brief pauses or silent intervals in vocal signals are common, and may become obscured by echoes which are present during the silent intervals. Echoes need not be discrete: sound is scattered off irregular surfaces (branches, leaves, grass) and will reach the listener by a variety of paths, and variability in path lengths will cause broadcast signals to be "smeared" in time. P.M. Waser and Brown (1986) employed RT60 to measure reverberation because it is a standardized index widely used in architectural acoustics. RT60 is the time it would take for a signal to decay 60 dB from its peak value following its offset. Using this index, reverberation was found to be prominent in both rain forest and riverine forest, but not in savanna (Fig. 4). Forest reverberation times were strongly influenced in both transmission distance and carrier frequency. Reverberation time in forested habitats for low-frequency tones (125 Hz) was nearly equal to that in the savanna. Evidently, sounds at this wavelength are not scattered much more in closed than in open habitats.

Environmental impediments to acoustic communication increase with propagation distance. Amplitude fluctuations and reverberation time increase with transmission distance in a way similar to that for attenuation. Hence, ecological constraints on the

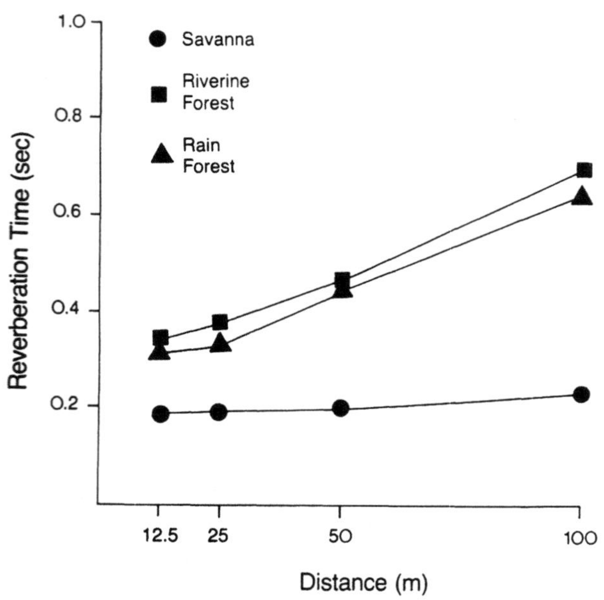

Fig. 4. Mean reverberation time (RT60) averaged across pulse-carrier frequencies in each habitat, as a function of broadcast distance (data from P.M. Waser and Brown 1986)

structure of communication sounds are uniformly more severe for long-range signals. All primate habitat types pose difficulties for amplitude and temporal modulation of signals, yet no studies quantitatively measure difficulties for frequency modulation. Future research should address the relative fidelity of frequency modulation as a function of transmission distance.

Altogether these results suggest that the elaboration of the fine structure of signals is likely restricted to short-range signals for species in all three habitat types. However, because the savanna habitat is free from problems created by reverberation, more elaboration and variation in fine structure may be expected in the signals used in this habitat. The availability of redundant information through the visual modality (Marler 1965) may supplement the perception of acoustic variants in open habitats. Thus, one may expect finer structural variation both between and within call classes in open-country primates relative to those seen in the vocal repertoire of closed-habitat primates. Though no quantitative cross-species data are available yet, the vocal patterns of the open-country vervets and baboons are suggestive. Vervet monkeys have a more diverse repertoire of short-range signals than do most of the forest *Cercopithecus* species (Gautier and Gautier-Hion 1977; Seyfarth and Cheney 1982), and baboons, though not quantitatively studied, have one of the richest and variable repertoires of short-distance calls, a repertoire which may be particularly well suited for revealing vocal tract configuration (Andrew 1963).

2.4 Habitat-Induced Distortion of Broadcast Vocalizations

Distortion, or changes in the fine structure of vocalizations, may occur independently in both the frequency or time domain. Nearly all natural vocalizations are composed of a family of frequency elements, and some of these are likely to be propagated less favorably than others. Thus the spectrum, or the frequency composition of an utterance, is likely to change or be distorted at progressively greater levels with increasing propagation distances. Complementing the likelihood of changes in the frequency composition of a call is the possibility of changes in the call's temporal pattern. Scattering may cause silent or low amplitude portions of signals to be overlaid by the indirectly transmitted wave fronts of louder components. Thus the temporal pattern as well as the overall duration of the signal may be changed by scattering and reverberation.

Not all signals, or all signal components, will be equally modified by reflection, absorption, and refraction. For example, transient elements of some signals may be immune to constructive or destructive interference if they are completed before wave fronts from the scattered indirect routes arrive. Similarly, since in a particular habitat disparate frequencies may attenuate at the same rate, signal elements composed of those frequency combinations will propagate without altering the ratio of the amplitudes of their components.

Because scattering and attenuation are independent processes, it is possible for signal degradation to occur independently in the time and frequency domains. That is, scattering may occur without differential absorption or differential attenuation of the frequency components of a signal. In such a case, a signal may have no change in the frequency domain, but may show degradation in the temporal patterning or time domain.

The converse may occur as well, where the signal's frequency components are differentially absorbed, yet the temporal envelope of the signal is preserved. Thus ideally measurements of signal degradation would be conducted in both domains.

Gish and Morton (1981) were the first to investigate, in Carolina wrens, the adaptation of calls to local habitat. To measure call degradation, they focused on the amplitude contour of broadcast signals, and measured the distribution of amplitude peaks in 3-dB bins and noted changes in the distribution of these peaks at a broadcast distance of 50 m. However, this index does not discriminate between the relative effects of echoes, constructive or destructive interference, or frequency-dependent attenuation. Furthermore, a signal could change markedly in both time and frequency domains, and still show no change in the distribution of amplitude peaks.

We thus describe here an alternative technique for the quantification of signal degradation in natural environments, and illustrate its use by describing the fidelity of blue monkey calls broadcast in Kenya's Kakamega forest. Broadcast signals were tape recorded on a Nagra IV-S recorder and a pair of AKG omnidirectional microphones (C452 EB/CK22) at transmission distances of 1, 12.5, 25, 50, and 100 m. Recordings were conducted in August, 1986, nominally the dry season, at an elevation of 8 m. Taped blue monkey calls, played back on a Nagra IV-L recorder and amplified by a 85 W Sony XM-700A amplifier, were broadcast from an Electro-voice Musicaster 100 loudspeaker at an elevation of 8 m. This system permitted the broadcast of very high fidelity, high amplitude signals. To assure representative measurements, signals were broadcast in three different directions for two repetitions, yielding a total of six measurements for each distance. Signals were calibrated as described in P.M. Waser and Brown (1986). Measurements of six representative blue monkey calls are described here: *double chirp, trill, grunt, boom, ka-train,* and *pyow* (Marler 1973). The first three calls are short-range calls, and the last three are long-range ones (Brown 1987).

Recorded signals were analyzed on an IBM AT computer utilizing ILS signal-processing software. Calls were digitized at a sampling rate of 27.5 kHz on a 12-bit A/D converter (Data Translation DT2818).

To obtain representative renditions of each signal at each propagation distance, the six samples of each call were digitally filtered, adjusted in amplitude to have the same power, and then aligned in time so that all the signals started at the same point in the files. The temporal alignment was performed by eye using time-expanded frames, and in difficult cases (unfavorable signal-to-noise ratios) by using cross-correlation techniques. Serially the amplitude of each data point was summed with the corresponding points in each file, and the total was divided by 6 to produce an averaged record. This approach yielded an average, or representative, signal in which each sample contributed an equal amount. That is, louder samples did not contribute more to the average than quieter ones. The alignment process is shown for two representative *boom* calls at broadcast distances of 1, 12.5, and 100 m in Fig. 5. As shown in this figure, the wave forms of the samples at 1 m are very similar, while those at 12.5 and 100 m are less consistent with one another. Average variations in the waveforms at these distances are shown in Fig. 6.

In order to characterize the average amount of change in the waveform of signals as a function of propagation distance, it is possible to calculate a normalized cross-correlation function for the averaged waveform at 1 m, and the corresponding wave-

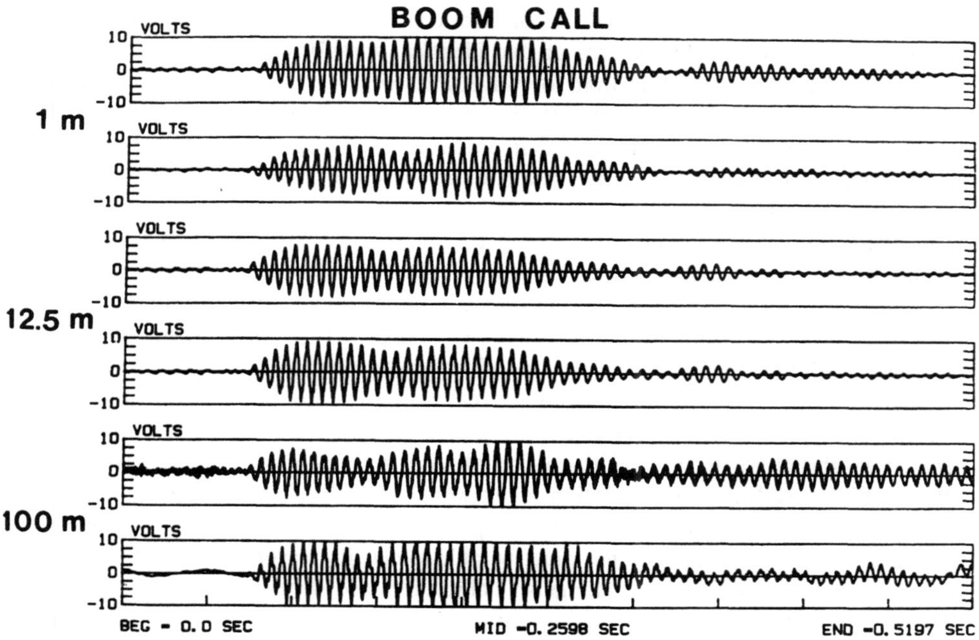

Fig. 5. Waveforms of two representative boom calls broadcast in Kakamega forest at 1, 12.5, and 100 m

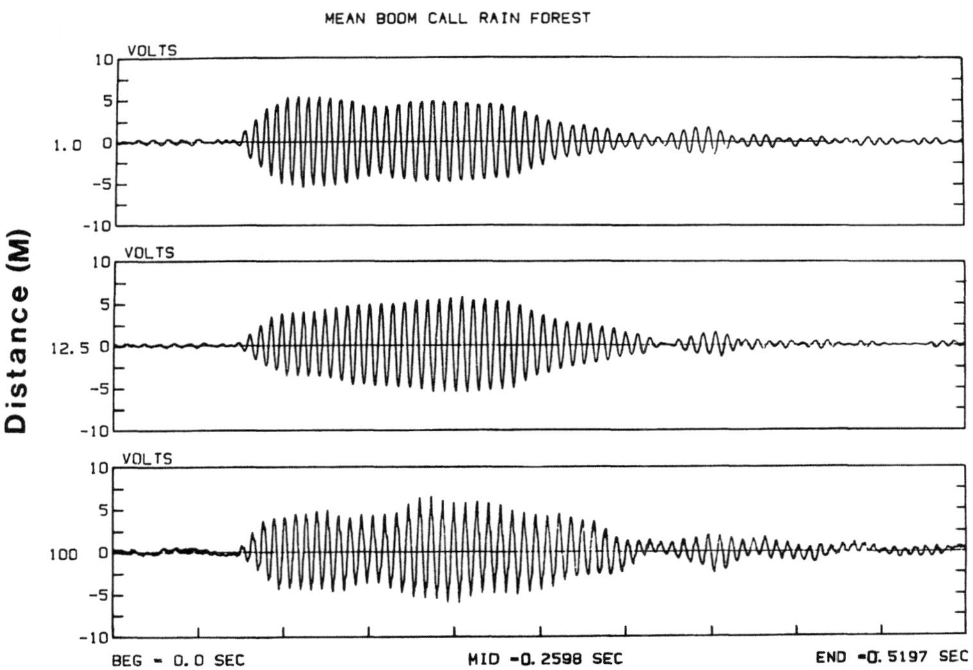

Fig. 6. Mean waveforms of the boom call at transmission distances of 1, 12.5, and 100 m

form at 12.5 or 100 m. The height of the ordinate for a normalized cross-correlagram ranges from 0.0 to 1.0. In the case of two identical waveforms the height of the correlagram would be 1.0, while that for waveforms which are progressively dissimilar, and hence uncorrelated, approach 0.0. We use correlagram height to characterize the magnitude of distortion in the time domain produced by propagation in natural habitats.

Table 1 displays habitat-induced distortion in the waveforms for all six calls at 12.5 and 100 m. As shown in Table 1 the correlation index ranges from 0.97 to 0.30. For five of the six calls the waveforms at 1 and 100 m are more dissimilar than are those at 1 and 12.5 m. In general, the three long-range calls, the *boom, ka-train,* and *pyow*, exhibit less distortion at 100 m (the average correlation index with the corresponding 1-m signal was 0.90) compared to the three short-range calls, *grunt, trill*, and *chirp* (the average correlation index with the corresponding 1-m signal was 0.82). These preliminary results suggest that the form of long-range calls is more resistant to distortion in the time domain than the waveforms of short-range calls. The *trill* signal was unusual in that it exhibited greater average distortion at 12.5 m (the correlation index was 0.30) than it did at 100 m (the correlation index was 0.81). The average waveforms for the trill at 1, 12.5, and 100 m are shown in Fig. 7. This magnitude of variability is quite dramatic and the auditory system must be able to classify waveforms with this degree of variability, as well as to distinguish habitat-induced distortion from potentially meaningful differences generated during production. These results suggest that while time-domain distortion tends to increase with transmission distance, this trend is not universal, and even at short transmission distances wave form distortion may be substantial.

To measure the corresponding possibility of distortion in the frequency domain the Fourier transform of the averaged waveforms for each signal is calculated and a difference spectrum is generated by subtracting the spectrum at 12.5 or 100 m from the spectrum of the corresponding waveform at 1 m. To examine changes in the relative spectra of signals, the power of the signals at 1, 12.5, and 100 m are first adjusted to be equal. Consequently, changes in average signal amplitude with transmission distance are controlled by this analysis. For example, Fig. 8 displays the *trill* call spectrum at 1 and 12.5 m, and the resulting difference spectrum. The difference spectrum displays both positive and negative values. Where the difference spectrum is negative, the relative amplitude of the signal at 12.5 (or 100 m) exceeds that at 1 m. Because of the inverse

Table 1. Time-domain change index: Normalized cross-correlation of the averaged signal at 1 m with the averaged signal at 12.5 or 100 m. The greater the difference between signals, the smaller the coefficient

Signal	12.5 m	100 m
Grunt	0.9667	0.8925
Pyow	0.9542	0.8571
Trill	0.3033	0.8063
Boom	0.9584	0.9413
Ka-train	0.9153	0.9101
Chirp	0.9348	0.7637

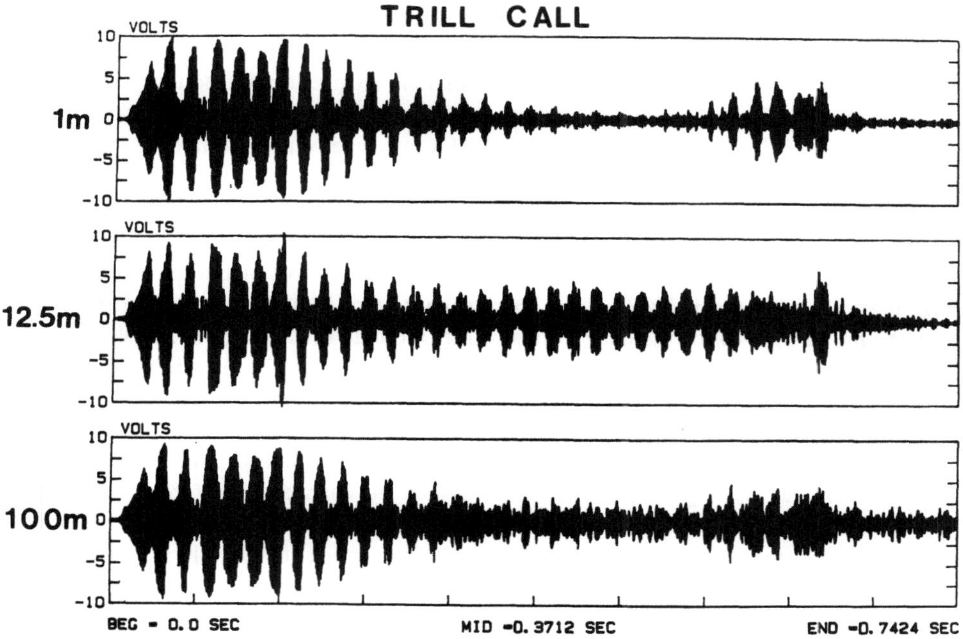

Fig. 7. Mean waveforms of the trill call at transmission distances of 1, 12.5, and 100 m

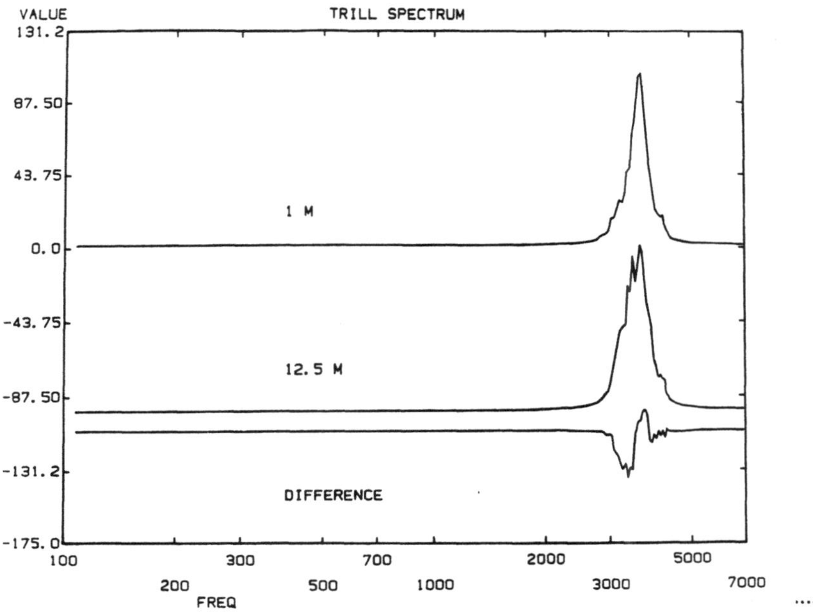

Fig. 8. Spectra of the trill call at transmission distances of 1 and 12.5 m in comparison with the difference spectrum of the two signals

square law, the absolute signal amplitude at distances greater than 1 m has almost certainly been reduced; however, some frequency components of the signal may be reduced less drastically than others, and the relative change in the frequency composition of the signal is shown by the difference spectrum. A difference spectrum denoted by a straight horizontal line would represent the case where the spectra of the signal at 1 m, 12.5, or 100 m were indistinguishable.

It is possible to derive a single number to represent the magnitude of the difference spectrum relative to that for the signal at 1 m. To do his we measure the absolute value of the area of the difference spectrum, and divide it by the area of the spectrum of the signal at 1 m. As seen in Table 2, the spectral change index at 100 m varied from 0.06 for the *boom* to 0.40 for the *chirp*. These values indicate that the *boom* signal changed minimally in the frequency domain as a result of transmission in the rain forest, while the *chirp* call exhibited marked change in frequency composition. The frequency change index did not vary in relationship with broadcast distance, however. In fact, only the *pyow* call showed a greater change in frequency composition at 100 m than it exhibited at 12.5 m. This finding suggests that habitat-induced changes in the frequency composition of vocalizations may occur close to the site of the signal's origin. Furthermore, this finding shows that the spectra of vocal signals may be substantially altered by frequency dependence in the processes of absorption, reflection, and refraction, calling into question the extent to which frequency-modulated signals are immune to environmental degradation (Wiley and Richards 1978; P.M. Waser and Brown 1984).

Does a signal's susceptibility to time-domain degradation correlate with its susceptibility to degradation in the frequency domain? Regression of time and frequency domain change indices yield correlation coefficients of 0.33 and 0.55 at 12.5 m and 100 m, respectively. Even at 100 m only 30% of the variation in one domain can be attributed to variation in the other. Signals like the *boom* appear to be relatively immune to degradation in either domain, while other signals may be more vulnerable to degradation in one domain compared to the other.

The potential significance of natural habitat acoustics on the structure of primate communication sounds has just begun to be examined. It seem likely that certain aspects of acoustic structure are relatively immune to degradation in any particular habitat, while others are not. If so, then signals may be expected to undergo selection, producing

Table 2. Frequency domain change index: Ratio of the difference in area of the power density of the averaged signal at 1 m and the averaged signal at 12.5 or 100 m. The greater the difference between signals, the larger the coefficient

Signal	12.5 m	100 m
Grunt	0.26	0.24
Pyow	0.20	0.35
Trill	0.34	0.11
Boom	0.10	0.06
Ka-train	0.34	0.14
Chirp	0.40	0.35

communicatively significant vocal contrasts composed of relatively degradation-immune elements. The characteristics of signals which influence their resistance to habitat-induced distortion have yet to be established for any species in any habitat. The techniques described here suggest how we might proceed.

Acknowledgments. We thank the Office of the President, Republic of Kenya for permission to pursue our fieldwork and the Kenyan Institute of Primate Research, and in particular Jim Else, for sponsorship and assistance. This research was supported by NIH grants RO1 NS16632-07 and KO4 NS00880-03.

References

Andrew RA (1963) Trends apparent in the evolution of vocalizations in the Old World monkeys and apes. Sym Zool Soc London 10:89–101

Bowman RI (1979) Adaptive morphology of song dialects in Darwin's finches. J Ornithol 120: 353–389

Brenowitz EA (1982) The active space of red-winged blackbird song. J Comp Physiol 147:511–522

Brown CH (1986) The perception of vocal signals by blue monkeys and grey-cheeked mangabeys. Exp Biol 45:145–165

Brown CH (1988) The active space of blue monkey and grey-cheeked mangabey vocalizations. Anim Behav

Brown CH, Schwagmeyer PL (1984) The vocal range of alarm calls in thirteen-lined ground squirrels. Z Tierpsychol 65:273–288

Brown CH, Waser PM (1984) Hearing and communication in blue monkeys *(Cercopithecus mitis)*. Anim Behav 32:66–75

Byrne RW (1981) Distance vocalizations of guinea baboons *(Papio papio)* in Senegal: an analysis of function. Behaviour 78:283–312

Chalmers NR (1968) The visual and vocal communication of free-living mangabeys in Uganda. Folia Primatol 9:258–280

Chappuis C (1971) Un example de l'influence du milieu sur les emissions vocales des oiseaux: l'evolution des chants en foret equatoriale. Terre Vie 25:183–202

Cosen SE, Falls JB (1984) A comparison of sound propagation and bird song frequency in marsh and grassland habitats. Behav Ecol Sociobiol 15:161–170

Gautier J-P, Gautier-Hion A (1977) Communication in Old World monkeys. In: Sebeok TA (ed) How animals communicate. Indiana Univ Press, Bloomington, pp 890–964

Gish SL, Morton ES (1981) Structural adaptations to local habitat acoustics in Carolina wren songs. Z Tierpsychol 56:74–84

Griffin DR (1971) The importance of atmospheric attenuation for the echolocation of bats *(Chiroptra)*. Anim Behav 19:55–61

Henwood K, Fabrick A (1979) A quantitative analysis of the dawn chorus: temporal selection for community optimization. Am Nat 114:260–274

Heuwinkel H (1982) Schalldruckpegel und Frequenzspektren der Gesänge von *Arcocephalus arundinaceus, A. scirpacues, A. schoenobaenus* und *A. palustris* und ihre Beziehung zur Biotopakustik. Oekol Voegel 4:85–174

Hunter ML, Krebs JR (1979) Geographic variation in the song of the great tit *(Parius major)* in relation to ecological factors. J Anim Ecol 48:759–786

Jilka A, Leisler B (1974) Die Einpassung dreier Rohrsängerarten *(Acrophalus schoenobaenus, A. scirpaceus, A. arundinaceus)* in ihre Lebensräume in bezug auf das Frequenzspektrum ihrer Reviergesänge. J Ornithol 115:192–212

Kristiansen UR, Pettersen OKO (1978) Experiments on the noise heard by humans when exposed to atmospheric winds. J Sound Vib 58:285–291

Marler P (1965) Communication in monkeys and apes. In: DeVore I (ed) Primate behavior: field studies of monkeys and apes. Holt, Rinehardt and Winston, New York, pp 554–584

Marler P (1967) Animal communication systems. Science 157:769–774
Marler P (1968) Aggregation and dispersal: two functions in primate communication. In: Jay P (ed) Primates: studies in adaptation and variability. Holt, Rinehart and Winston, New York, pp 420–438
Marler P (1973) A comparison of vocalizations of red-tailed monkeys and blue monkeys, *Cercopithecus ascanius* and *C. mitis* in Uganda. Z Tierpsychol 33:223–247
Marten K, Marler P (1977) Sound transmission and its significance for animal vocalization I: temperate habitats. Behav Ecol Sociobiol 2:271–290
Marten K, Quine D, Marler P (1977) Sound transmission and its significance for animal vocalization II: tropical forest habitats. Behav Ecol Sociobiol 2:291–302
McGregor PK, Falls JB (1984) The response of western meadowlarks *(Sturnella neglecta)* to the playback of undegraded and degraded songs. Can J Zool 62:2125–2128
McGregor PK, Krebs JR (1984) Sound degradation as a distance cue in great tit *(Parus major)* song. Behav Ecol Sociobiol 16:49–56
Morton ES (1970) Ecological sources of selection on avian sounds. Ph D Diss, Yale Univ
Morton ES (1975) Ecological sources of selection on avian sounds. Am Nat 109:17–34
Nottebohm F (1975) Continental patterns of song variability in *Zonotrichia capensis*: some possible ecological correlates. Am Nat 109:605–624
Payne RS, Guinee LN (1983) Humpback whale *(Megaptera novaeangliae)* songs as an indicator of "stocks". In: Payne R (ed) Communication and behavior of whales. Westview, Boulder, pp 333–358
Payne RS, Webb D (1971) Orientation by means of long range acoustic signaling in baleen whales. Ann NY Acad Sci, pp 110–141
Potash LM (1972) A signal detection problem and possible solution in Japanese quail. Anim Behav 20:192–195
Richards DG, Wiley RH (1980) Reverberations and amplitude fluctuations in the propagation of sound in the forest: implications for animal communication. Am Nat 115:381–399
Ryan MJ (1986) Environmental bioacoustics: evaluation of a commonly-used experimental design. Anim Behav 34:931–933
Ryan MJ, Brenowitz EA (1985) The role of body size, phylogeny, and ambient noise in the evolution of bird song. Am Nat 126:136–139
Seyfarth RM, Cheney DL (1982) How monkeys see the world: a review of recent research on East African vervet monkeys. In: Snowdon CT, Brown CH, Petersen MR (eds) Primate communication. Cambridge Univ Press, Cambridge, pp 239–252
Shy E (1983) The relation of geographical variation in song to habitat characteristics and body size in North American tanagers (Thraupinae: *Piranga*). Behav Ecol Sociobiol 12:71–76
Waser PM (1982) The evolution of male loud calls among mangabeys and baboons. In: Snowdon CT, Brown CH, Petersen MR (eds) Primate communication. Cambridge Univ Press, Cambridge, pp 117–143
Waser PM, Brown CH (1984) Is there a sound window for primate communication? Behav Ecol Sociobiol 15:73–76
Waser PM, Brown CH (1986) Habitat acoustics and primate communication. Am J Primatol 10: 135–154
Waser PM, Waser MS (1977) Experimental studies of primate vocalizations: specializations for long-distance propagation. Z Tierpsychol 43:239–263
Wasserman FE (1979) The relationship between habitat and song in the white-crowned sparrow. Condor 81:824–826
Webb DC, Tucker MJ (1970) Transmission characteristics of the sofar channel. J Acoust Soc Am 48:767–769
Wiley RH, Richards DG (1978) Physical constraints on acoustic communication in the atmosphere: implications for the evolution of animal vocalization. Behav Ecol Sociobiol 3:69–94
Wiley RH, Richards DG (1982) Adaptations for acoustic communication in birds: sound transmission and signal detection. In: Kroodsma DE, Miller DE, Miller EH (eds) Acoustic communication in birds, vol I. Academic Press, London New York, pp 131–181

Social Interactions

A New Approach to Finding Components Essential for Intraspecific Communication

Manfred Maurus, Klaus M. Streit, David Barclay, Elisabeth Wiesner, and Brigitte Kühlmorgen [1]

Understanding primate vocal communication requires knowledge of which utterances are signals for the conspecific and how they influence his behavior. A most difficult part of endeavors to gain such knowledge is that of finding the "natural" categories of the vocal repertoire, i.e. those vocal structures which are relevant for the species under investigation. In analyzing the monkeys' communication, the observer is confronted with an enormous variety of acoustic structures and use of these structures by the monkeys which he very often does not understand. If an observer can base his categorization on structural properties only, even the most refined methods of structural analysis, e.g. highly sophisticated computer analyses, do not yield "natural" categories, for functional relations cannot be taken into account. Observer intuition inspired by listening to the monkeys' calls or by studying sonagrams does not suffice: recognizing call types by listening is a process depending on the observer's physical abilities to differentiate the acoustic structures and his propensities for categorizing sounds, capacities varying from observer to observer and varying at most from the observer to the animal under investigation. "The spectrogram can show in fine physical detail which sounds have been produced but cannot show which features of the sound are communicatively significant for the species or for the particular animal community." (Hockett 1960). Moreover, we even do not know how much the degree of differentiation available from sonagrams corresponds to the degree of differentiation in the acoustic perception by the monkeys. When taking these considerations into account, it is not surprising that different authors produce different categorizations for the same species [e.g. Reynolds (1976) for rhesus monkeys, Newman (1985) for squirrel monkeys].

When studying vocal communication processes, the observer can record the vocalizations uttered and is then capable of analyzing the structural characteristics. The animals' intraspecific communication, however, is based on the functional characteristics of the signals used. The interrelationship between functional and structural properties is a priori unknown to the observer. The units obtained by categorizing form the basis for further investigations. All calculations, deductions and conclusions based on these units are wrong if only some of the units are defined incorrectly. Hence correct categorization is a very important prerequisite for many investigations of primate vocal communication. If the observer wants to know which structures are essential to which function, he can find out the interrelationships between function and structure by applying a structural analysis and using his knowledge of function. The following examples may illustrate this.

1 Max Planck-Institut für Psychiatrie, Kraepelinstr. 2, 8000 München 40, FRG

Todt/Goedeking/Symmes (Eds.)
Primate Vocal Communication

In the Japanese monkeys' "coo" calls, Green (1975), by "refining the concept of what is a distinctive sound and reexamining the notions of what constitutes a similarity in behavioral circumstances", could identify structural components contained in the "coos" which were linked to different behavioral circumstances. Each of the structural components he separated corresponded to behavioral circumstances from which he could deduce the function of the structural characteristics under "refined investigation".

The vervet monkey's "grunt" is uttered under different social circumstances, indicating that this call may have different functions. Skillful experimentation showed, however, that there are four calls differently produced by the monkeys and used in four different social circumstances (Cheney and Seyfarth 1982). Although the observers, neither by listening nor by evaluating sonagrams (Seyfarth and Cheney 1984), could discriminate between the four calls, in principle it is possible to detect the structural components responsible for functional differences. It seems to be a question of finding an appropriate method.

Masataka (1983) selected several call components of Goeldi's monkeys and modified each of them. He found one component (the amplitude sweep) which, if modified, induced a categorical differentiation in the conspecific's response. The functional boundary (deduced from the categorical differentiation in the conspecific's response) clearly corresponds with a structural boundary in the component concerned.

In the examples reported above, the function of the vocalization became evident by considering the social circumstances to which the calls are linked. Thus the knowledge of the function became the basis of identifying the structural component linked with the function. The interrelation between structure and function could be analyzed because there was an indication about function. However, there is also vocal interaction under circumstances not indicative of function (e.g. if a vocalization of unknown function precedes, occurs simultaneously, or succeeds vocal, visual and/or tactile behavior of unknown function). Here neither the structural patterns relevant to communication are known nor can a functional differentiation be derived from the accompanying behavior or circumstances. Hence, for the problem of finding relations between structure and function, the only approach possible (in this situation) is to start with structurally analyzing the vocalizations produced by the monkeys and then testing the functional relevance by asking the monkeys themselves. This approach is founded on two main procedures: the first is to explore *the entire* vocal repertoire of the species to find structural patterns that may be relevant for intraspecific communication; the second procedure is to examine whether these structural patterns have a communicative function by testing their influence on the conspecific's succeeding vocal behavior.

1 Structural Analyses

This procedure begins by splitting the calls of the repertoire into small elements. In this step, the decision where to split and which quantities to use for describing the elements may be a source of observer-induced errors. However, the finer the repertoire is split, the smaller is the danger of bias. The elements are described by frequency, amplitude and time measures and quantities derived thereof.

It seems reasonable to believe that monkeys produce components used for transmitting information more precisely than those not transmitting information. As far as discrete units are concerned, this will result in an accumulation of the number of occurrence at particular regions of the scales of the relevant parameters, a principle realized in human language (e.g. VOT: Zlatin 1974; Sweeting and Baken 1982) and in monkey communication as well (e.g. Masataka 1983). Relying upon this consideration, we looked for accumulations in the quantities used for describing the elements gained by splitting. Cluster analyses produced classes in which the elements of one class were similar in respect to one or more of the quantities considered and dissimilar to elements belonging to another class (for more details see Maurus et al. 1984). These classes represent structural components which may or may not be essential for communication. They are the subject of the succeeding analytical procedure.

2 Functional Analyses

The only proof of a component being essential for intraspecific communication is to unambiguously demonstrate that it influences the conspecific's behavior. This influence cannot simply be proven by using playbacks, for instance, and examining the animal's "response". Even if we are "sure" to have a "response" but the communicative function of it is unknown, a situation occurring most frequently in the behavior we are investigating, we are still unable to recognize whether or how the conspecific's behavior has been influenced. Due to this uncertainty, and as long as we are not able to deal with the enormous variability of the conspecific's "responses", even large statistical samples will not help.

To define the behavioral or situational categories by intuitive interpretation in order to establish correlations between structural components and elicited behavioral "responses" would be incorrect, for such an interpretation would anticipate results. Defining results as prerequisites for subsequent analysis instead of gaining them by the analysis of unbiased data is circular reasoning.

For an approach aimed at avoiding such errors the anticipation of a functional relationship is unacceptable. On the other hand, at least two functionally different categories are required for measuring the effect a component has on the conspecific. This difference cannot be stated arbitrarily. It has to be proven. A noncircular categorization under these circumstances must be based on structural quantities. Hence, firstly, the two categories must be separated unambiguously with regard to structural properties exclusively. Secondly, the categories established in this way must be proven to be functionally different.

In order to fulfill the first requirement, we divided the entire vocal repertoire of squirrel monkeys into two categories. We labeled them x and z (Fig. 1) and assigned all calls with at least one component showing a frequency modulation of more than 500 Hz between the lowest and highest frequency or with a fundamental frequency higher than 3 kHz to category z; all calls not possessing these characteristics were assigned to category x. More casually, "tonal" calls (without as well as with "noise" components) were assigned to category z, while "noisy" calls (containing no "tonal" components or "tonal" components with a frequency modulation of less than 500 Hz and

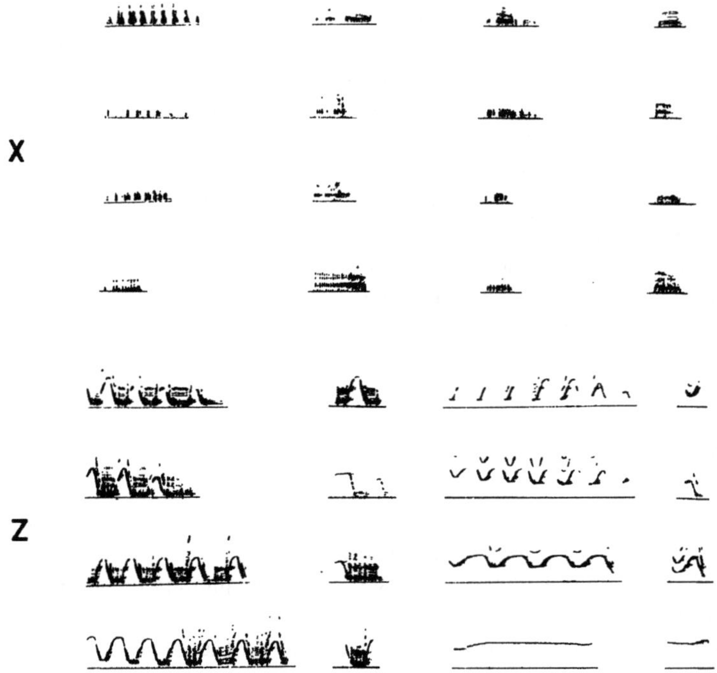

Fig. 1. Sonagrams demonstrating some calls of category *x* (*above*) and category *z* (*below*)

F_ϕ below 3 kHz) were assigned to category x. The assignments were based on the checks and cross-checks of up to five observers who had listened to the calls or to manually defined time frames from the recording, and studied the oscillogram and a coupled spectrographic on-line display (for details see Streit et al., in prep.). The rate of doubtful classification was far below 1 ‰ of the number of calls classified. Calls as to which there were doubts were differently coded and by no means alter the results.

In order to fulfill the second requirement, we selected one monkey as test animal T_1 and placed it in a cage where it was joined consecutively for maximally 80 min per day by one of its partners P_1–P_5. This set of experiments was repeated several times with the succession of partners being randomized (Fig. 3). The whole procedure was repeated with another animal as test animal T_2 and its partners P_2–P_5 being the same as above. With T_2 and its partners P_2–P_5 an additional set of experiments was executed in which visual, tactile and olfactory contact between T_2 and its partners P_2–P_5 was excluded by experimental manipulations (Maurus et al. 1985, 1987).

T_1 and T_2 as well as P_1, P_2, P_3 and P_4 were adult Gothic-arch males, P_5 was an adult Gothic-arch female. All animals had been kept in captivity for several years in breeding groups consisting of three to six animals. During experimental periods the animals P_1–P_5 were kept alone in single cages (3 m^3 in size) where they could hear but not see each other. T_1 and T_2, respectively, were permanently kept in an experimental cage of about 4.6 m^3 as described in Maurus (1974, Fig. 4). During the experiments the animals (T and P) were fed normally. No relevant stimuli from outside the cage, acoustic as well as visual ones, were given to the animals. The calls of both animals were tape-recorded telemetrically (Maurus and Szabolcs 1971) over separate channels, and the time (in ms) of each vocal event was registered by a time-code generator. Visually recognizable behavior was filmed and timed. The vocalizations of T and P were then arranged in a list according to their starting time [Fig. 2, where P's utterances are parenthesized: (x), (z)]. This list showed periods with only one of the monkeys vocalizing ("monologues") and other periods with both monkeys vocalizing ("dialogues"). By comparing intraindividual transitions in T's utterances during monologues with those during dialogues, we found that there was much less influence by the different partners on T's transition rates when T was monologuing (P silent) than in dialogues (P vocal). This result indicates that T's transition rates of x and z calls depend extremely little on which partner is present (left column, Fig. 3) but on what or how each of the partners is vocalizing (right column, Fig. 3).

If we compare, for instance, the experimental situation of T being in the cage together with P_4 with that of T being together with P_5, we find that T's intraindividual transitions are very similar as long as P_4 or P_5 keep silent. When P_4 or P_5 are vocalizing, the intraindividual transitions of T are interrupted by vocalizations of the P's. If T's intraindividual transition rates are different when P_4 vocalizes (right column, Fig. 3) from T's intraindividual transition rates when P_4 is silent (left column, Fig. 3), it indicates that P_4's vocalizations have a certain influence on T's vocalizations. If the same happens when P_5 is the partner and, additionally, the change in T's intraindividual transition rates is different from that when P_4 vocalized, it suggests that it is not the individual's presence that is important but his/her vocalizations (as T's transition rates differ much less when partners are silent). This suggestion is supported by the data gained from experiments excluding visual, tactile and olfactory contacts between T and P animals (Maurus et al. 1987).

Still unclear, however, is the type of influence exercised by P's utterances: is it the fact *that* P vocalizes, *what* it vocalizes or *how* it vocalizes? In the case of *what*, acoustic

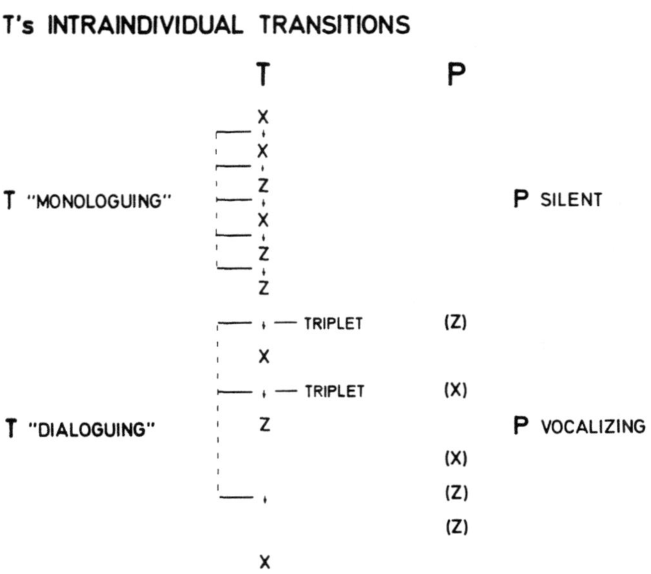

Fig. 2. Schematic drawing of the lists of vocalizations from *T* and *P* showing T's intraindividual transitions (↓) in periods connected by *dashed lines*: while *P* is silent ("monologues"), and while *P* is vocalizing ("dialogues"). Also shown are the special intraindividual transitions interrupted by exactly one call of *P*: triplet ("... while ... silent [vocal]" means that between the starts of two consecutive calls of one animal the other animal uttered no call [at least one call])

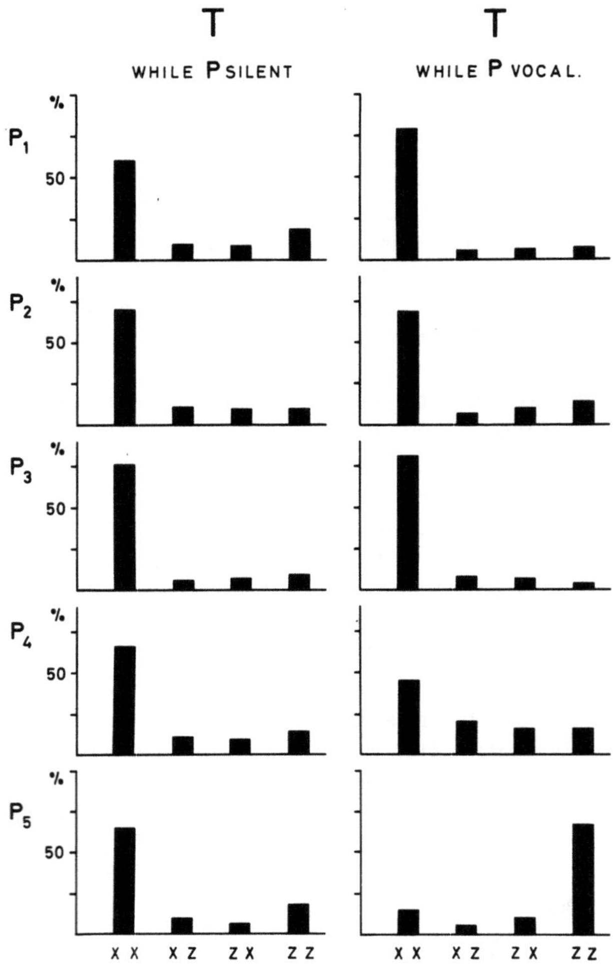

Fig. 3. Relative frequencies of intraindividual transitions uttered by *T* while P_1–P_5 were silent (*left column*) and while P_1–P_5 vocalized (*right column*) ("... while ... silent [vocal] means that between the starts of two consecutive calls of one animal the other animal uttered no call [at least one call])

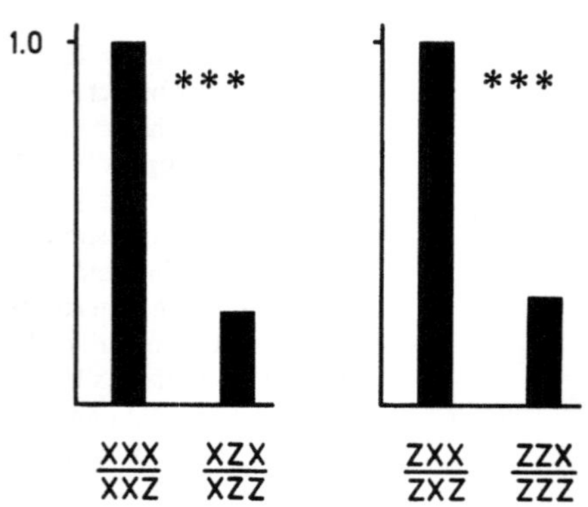

Fig. 4. Ratios of intraindividual *xx* transitions to *xz* transitions in triplets when the conspecific utters an *x* in between (*first column from left*) or a *z* (*second column from left*) and ratios of intraindividual *zx* transitions to *zz* transitions in triplets when the conspecific utters an *x* (*third column from left*) or a *z* (*fourth column from left*). Significance test (χ^2 test, d.f. = 1): *** = $p < 0.0000$; n = 6.077 triplets

structures are involved which transfer more information to the recipient than merely sender identification. Consequently, those structures influence the conspecific's behavior regardless of who the sender is, and who the recipient. By *how* we mean acoustic structures characteristic of a particular animal which may inform the recipient of who the sender is, regardless of the structures that may transfer information about *what*.

Supposing such theoretical separation is reasonable and, although the *what* components may interfere with the *how* components, for a start it seems plausible to investigate each of them independently, beginning with the *what* structures. These structures may be expected to influence the conspecific's behavior regardless of who the T animal, who the P animal, who the sender and who the recipient is (for a more detailed discussion see p. 83). As such influence must become evident by pooled data, we studied all that gained from experiments involving the animals T_1, T_2 as well as P_1-P_5, concentrating on all those instances when, between two succeeding calls of one animal (intraindividual transitions), exactly one call was uttered by another animal (triplets; Fig. 2). Measuring the influence that a call has on the conspecific's succeeding vocal behavior is based on the following considerations.

All intraindividual transitions beginning with an x call and interrupted by a conspecific's x call can end either with an x, thus forming an xxx triplet, or with a z call, forming an xxz triplet. Counting all xxx triplets and dividing them by the number of all xxz triplets, $\frac{xxx}{xxz}$, yields a measure for the relation of xx intraindividual transitions to the xz intraindividual transitions when they are interrupted by the conspecific's x call. This same procedure may be applied when these transitions are interrupted by a z call instead of an x call. It would yield a number of xzx triplets that can be divided by the number of xzz triplets for calculation of their relation: $\frac{xzx}{xzz}$. If the value of $\frac{xxx}{xxz}$ is significantly different from that of $\frac{xzx}{xzz}$, it indicates that the conspecific's x calls, when interrupting intraindividual transitions, exercise an influence that differs from that exercised by z calls interrupting intraindividual transitions. All intraindividual transitions beginning with a z call, $\frac{zxx}{zxz}$ as compared to $\frac{zzx}{zzz}$, can also be examined this way. The results in Fig. 4 show highly significant differences between $\frac{xxx}{xxz}$ and $\frac{xzx}{xzz}$ as well as between $\frac{zxx}{zxz}$ and $\frac{zzx}{zzz}$. This is proof that in structural category x there are calls functionally (influencing the conspecific's behavior) different from calls in structural category z (for more details see Maurus et al. 1985).

This means that by structurally categorizing the entire vocal repertoire in categories x and z, we arrived at a functional dissimilarity unprejudiced by observer decision in the functional properties of the call. Although this functional discrimination is a rough one, in conjunction with intraindividual transitions in triplets, it is a useful means for identifying the communicative significance of vocal components.

For testing the influence a component x_i, for instance, has on an animal's intraindividual transition, we can compare the $\frac{xx_ix}{xx_iz}$ ratio versus the $\frac{xxx}{xxz}$ ratio. $\frac{xxx}{xxz}$ includes all triplets with an x between the intraindividual transitions, i.e. all x_i and all x-x_i (thus denoting all x, not x_i). If $\frac{xx_ix}{xx_iz}$ differs significantly from $\frac{xxx}{xxz}$, x_i calls, influence on the subsequent vocal behavior differs from that of all x including x_i (expectation value). In the same way we can compare the influence that x_i has with that influence which the x-x_i have or with that influence another component being contained in x-x_i has, for instance x_j. If $\frac{xx_jx}{xx_jz}$ is equal to $\frac{xx_ix}{xx_iz}$, the components x_i and x_j do not differently influence the succeeding vocal behavior. They may or may not be differently perceived, but are not differently responded to. But if $\frac{xx_ix}{xx_iz}$ differs significantly from $\frac{xx_jx}{xx_jz}$, x_i and x_j are differently responded to and therefore differently perceived. They are part of the "natural" categorization of the vocal repertoire. If an x_i has been proven to be functionally different from an x_j, the comparatively rough measure of $\frac{..x}{..z}$ can be substituted by $\frac{..x_i}{..x_j}$, a procedure which may eventually lead to more refined results. This test can be executed for each component regardless of whether frequency-, time- and/or amplitude-related (e.g. power spectra) (Streit et al., in prep.) or frequency-modulated (Barclay et al. 1987). In the following, its application will be demonstrated by considering some structural properties of amplitude-modulated calls.

3 Amplitude Leaps

The calls of category x show no or only small frequency modulation. If there are essential components at all in the structural diversity of this category, they may be associated with amplitude or frequency distribution patterns rather than with frequency modulation. We use amplitude patterns, more exactly the amplitude modulation of x calls, as an example to demonstrate structural as well as functional analyses.

For each x call, we calculated the envelope of its amplitude time course. Each minimum in this course is taken for splitting the call into segments. By doing so, we obtain call segments with minimal amplitudes at both ends of the segment and maximal amplitude in between. Each segment can be described by a series of quantities comprising amplitude and time measures (for more details see Maurus et al. 1984). Each call can then be described by its segments. Paying attention to the difference between the amplitude maxima of the segments succeeding each other in a call, we can calculate the quotient: maximal amplitude of the higher segment divided by the maximum amplitude of the lower segment for each pair of immediately succeeding segments. Most often this quotient is very small. That means that most segments succeeding each other do not vary significantly in amplitude. Medium values are seldom, while high values frequently occur again. Thus, the number of occurrences is not evenly distributed on the scale of amplitude differences but is accumulated at both ends (Fig. 5). The accumulation of minor differences seems to be trivial because it indicates that there are many segments succeeding each other, having only a small deviation in amplitude maxima. But in conjunction with the decrease in the number of medium differences and the increase of major differences, the distribution seems to indicate a categorical differentiation at least as far as call production is concerned. This production of categorical amplitude differences may be due to a communicative function of amplitude modulation. There are calls with almost constant amplitude for their entire duration

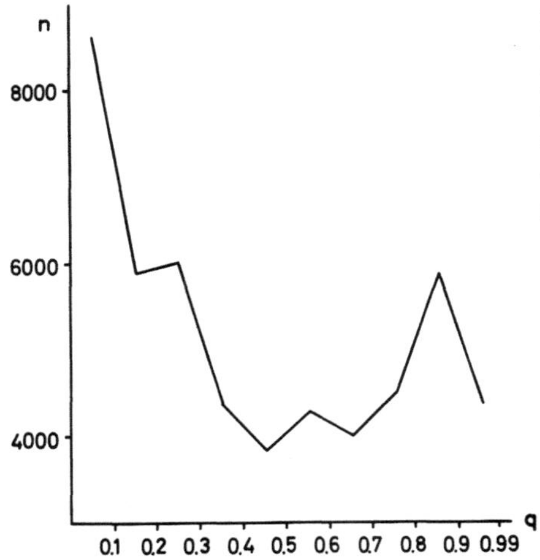

Fig. 5. Histogram of quotients q between the amplitude maxima of immediately adjoining segments. Data from x calls uttered by five animals. (The upward difference to the first segment of the call and the downward difference from the last segment of the call were not included, hence excluding differences with value 1.0)

and calls with one or more major amplitude changes, upward leaps as well as downward leaps, at different locations within the call.

What makes the amplitude leap a component essential for communication? Is it the fact that it exists within a call or not? Is it the number of occurrences within a call? Is it the position within a call? Is it the upward or downward leap? Or is it a more complex constellation of these properties? We found that the number of amplitude leaps produced per call depends on the social situation to which the monkey is exposed (Maurus et al. 1984, 1986). This is another hint at a communicative function.

Furthermore, some calls characterized by different types of amplitude modulation were presented to squirrel monkeys. Single-cell recordings in parts of the auditory pathway showed neural activity necessary for perception of amplitude leaps as well as their number in the call (Müller-Preuß and Maurus 1985; for more details see Müller-Preuß, this Vol.).

At this point we have many hints but no convincing proof that amplitude modulation affects the conspecific's behavior, nor do we know which of the manifold constellations of amplitude changes in the call are essential. In order to answer this question, we can make use of the intraindividual transition influenced by conspecifics' vocalizations and try to analyze which of the many amplitude leap positions possible within x calls are differently responded to by the animals and which are not.

In the example to be demonstrated we referred to those amplitude leaps only in which the maximal amplitude of the succeeding segment is at least five times higher than the maximal amplitude of the preceding segment in upward leaps and at least five times smaller in downward leaps. Considering only these upward and downward leaps and disregarding all structural properties that the calls additionally may have we can ask, is it the fact that a call has amplitude leaps, or is it the arrangement of the amplitude leaps within a call which is responsible for influencing the conspecific's behavior? If we compare all x calls that have no amplitude leaps with those containing at least one (upward or downward), our data, derived from 3.045 triplets, do not indicate a significant difference ($p = 0.2$).

In addition, we can subdivide the x calls containing amplitude leaps into six classes to arrive at seven classes of x calls:

1st class: all x calls not containing any amplitude leaps (–);
2nd class: all x calls containing exactly one upward amplitude leap (↑);
3rd class: all x calls containing exactly one downward amplitude leap (↓);
4th class: all x calls in which the first amplitude leap is an upward leap and the last one in the call a downward leap (↑↓);
5th class: all x calls in which the first amplitude leap is a downward leap and the last one in the call an upward leap (↓↑);
6th class: all x calls in which the first as well as the last leap is an upward one (↑↑);
7th class: all x calls in which the first as well as the last leap is a downward one (↓↓).

Table 1 indicates that calls of class 5 influence the succeeding vocal behavior most differently from calls containing none or only one amplitude leap upwards as well as downwards. There is also a difference between calls having one downward amplitude leap and those where the first amplitude leap goes up and the last one down. From the data contributing to Table 1 it can be deduced that:

1. Calls containing exactly one upward amplitude leap differ significantly ($p < 0.05$) in influencing the succeeding vocal behavior from the calls having one upward plus at least another upward or downward leap;
2. There is a significant difference ($p < 0.05$) between calls with exactly one downward leap and calls with one downward leap plus at least another downward or upward leap;
3. There is a significant difference ($p < 0.05$) between all x calls containing exactly one amplitude leap (upward or downward) and all x calls containing more than one leap (upward or downward).

Table 1. Pairwise comparison of the influence of call components 1–7 on succeeding vocal behavior[a]

	1	2	3	4	5	6	7	
1		■■			**		■	–
2	1902				*		■■	↑
3	1866	618		*	**		■	↓
4	1755	507	471			■		↑↓
5	1947	699	663	552			*	↓↑
6	1710	462	426	315	507			↑↑
7	1740	492	456	345	537	300		↓↓
	–	↑	↓	↑↓	↓↑	↑↑	↓↓	

[a] Upper triangle: pairs significantly different. Significance test (χ^2 test d.f. = 1): * ($p < 0.05$), ** ($p < 0.005$); also most similar pairs are shown: ■ ($p > 0.6$), ■■ ($p > 0.85$). Lower triangle: number of triplets in each sample.

With the instrument presented here, it can only be proven that there is a difference in function between two or more structurally defined call classes. Neither the kind of function nor the functional characteristics that distinguish one from the other are indicated by it. Although this may seem to be a very poor result, it is an indispensable step in the search for the structural boundaries related to the functional classes within the vocal repertoire. This procedure has to be extended to all physical parameters forming call structures; two briefly outlined examples follow. How to proceed in analyzing functional interrelationships of the call classes will be treated in the Discussion.

4 Power Spectra

With respect to the role of amplitude leaps in squirrel monkey communication demonstrated above and the possible role of the power distribution (or its changes) (Cheney and Seyfarth 1982; Seyfarth and Cheney 1984), the question arises whether there are interrelations between amplitude and frequency parameters.

For an analysis of the communicative function of frequency components in x calls, we calculated a power spectrum of each segment of all x calls (using the same process of segmentation described above). From each spectrum we selected the frequency of the most prominent peak (labeled MA1MF, see Fig. 6) for further analysis (detailed description of calculation techniques in Streit et al., in prep.). In Fig. 7 a histogram is plotted showing the distribution of MA1MF values. It shows four maxima which are

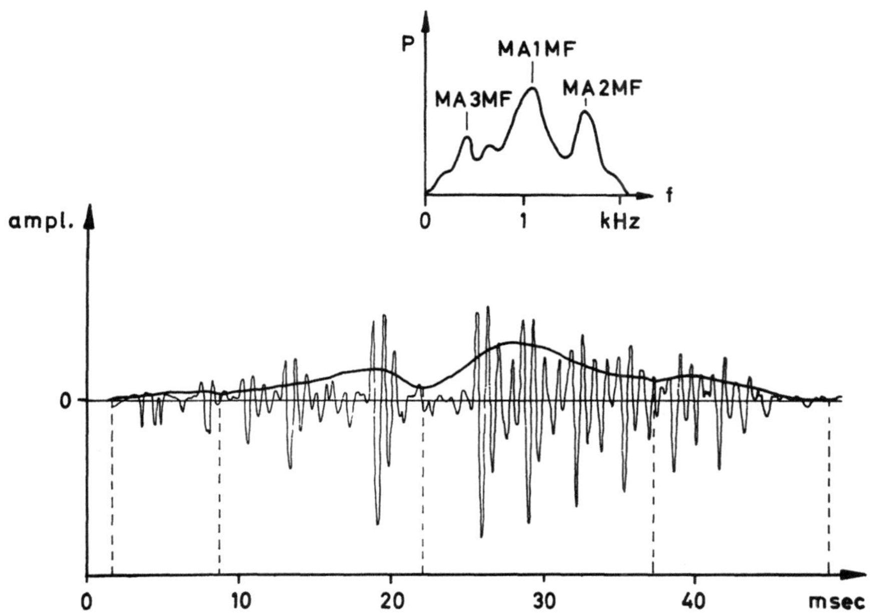

Fig. 6. *Below:* oscillogram of an x call and its envelope of the amplitude. The limits of segments defined by the minima of the envelope are drawn as *dashed lines. Above:* power spectrum of the segment defined below. The frequencies of the three most prominent peaks are marked as *MA1MF*, *MA2MF* and *MA3MF*

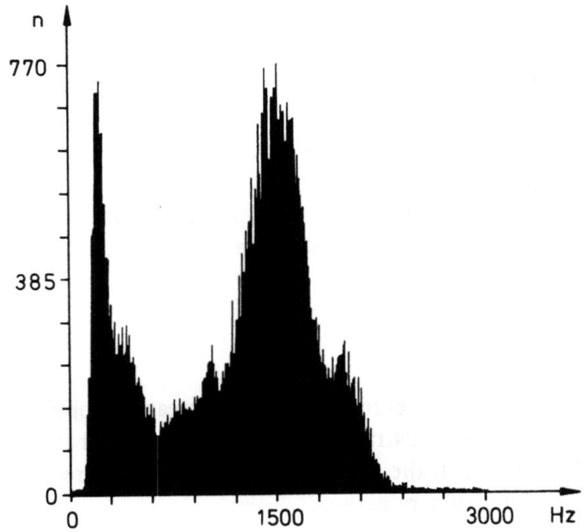

Fig. 7. Histogram showing the distribution of MA1MF values. *Abscissa:* values of MA1MF up to 3 kHz; *ordinate:* number of cases of MA1MF values. The histogram includes the values of 68465 segments. There were 319 segments with an MA1MF value of more than 3 kHz. The frequency bands defined for structural analysis are delimited at 700, 1100 and 1700 Hz

separated by minima at 700, 1100 and 1700 Hz. For further investigation we divided the frequency scale into four bands according to these minima.

A frequency change shall be defined here as a change of MA1MF from one band to another when proceeding from one segment to the next. Analysis of amplitude leaps revealed that the direction of changes in amplitude is not strictly coupled with the direction of changes in frequency. Calls containing exactly one downward amplitude leap with a downward change in frequency had a significantly different influence on the succeeding vocalization than vocalizations containing exactly one downward amplitude leap with an upward change or no frequency change (χ^2 test d.f. = 1, $p <$ 0.05). This proves that the direction of frequency changes has an additional communicative function for calls with a downward amplitude leap.

5 Frequency Modulation

Z calls were separately spectrographed at half speed with a Kay sonagraph 6061-B and the time course of the fundamental frequency digitized via a digitizing tablet (25.01 ms cm^{-1}, 3333.33 Hz cm^{-1}; 1 tablet unit = 0.1 mm). The complexity of frequency modulation over time and different call lengths make parameterization difficult. Hence, the contours thus obtained were segmented using a linear approximation algorithm with nonfixed boundaries (Barclay et al. 1987). Three quantities were found to carry most of the information pertaining to such segment approximation and thus judged to be appropriate for segment description: the y coordinate of the center of gravity, slope and segment length (duration). By discretizing these quantities, each call could then be represented by a sequence of symbols; for structural analysis so-called syntactic methods were applied (see Fig. 8).

Based on accumulations in the frequency of occurrence of parts of such symbol sequels, z calls were further classified into two categories. The first category z_A in-

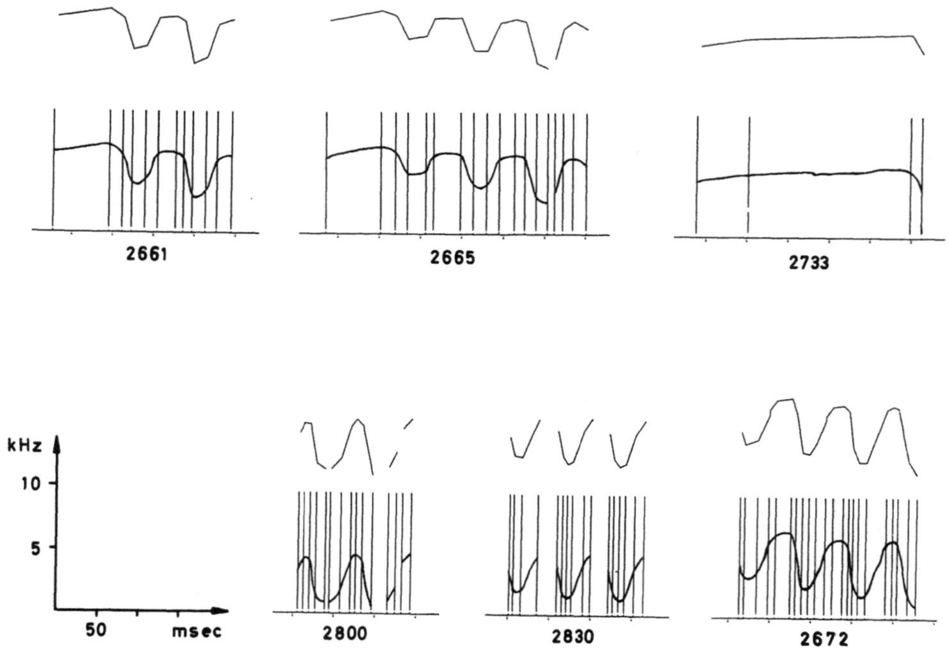

Fig. 8. F_0 contours of some z calls: *vertical lines* show their segmentation points resulting from a split-and-merge procedure thus obtaining the linear approximation above each call

cludes all calls containing (1) one segment very low in frequency and with a steep slope or (2) one segment very low in frequency but with only a moderate slope, if connected with an interruption in the contour as available from the sonagram.

The second category z_B contains all the remaining z calls. Call numbers 2800 and 2672 in Fig. 8 are z_A calls.

The result of testing these components for functionality is shown in Fig. 9: z_A and z_B calls, when preceded by a z call, influence vocal behavior of the partner animal quite differently.

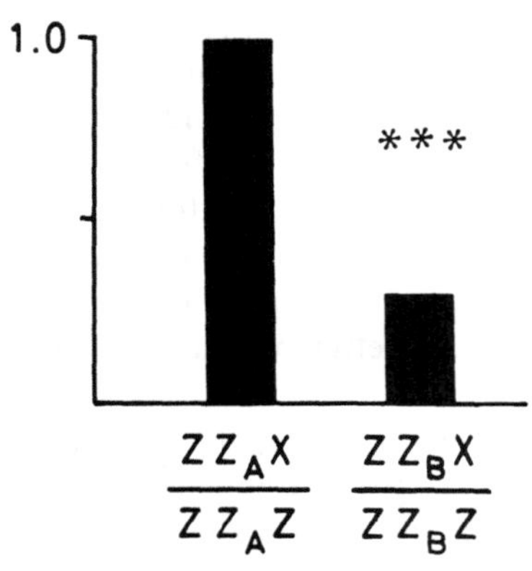

Fig. 9. Influence of z_A and z_B calls on the subsequent vocal utterance when z_A or z_B were preceded by a z call ($n = 561; p < 0.0000, \chi^2$ test, d.f. = 1)

6 Discussion

So far we have proven that amplitude modulation, power spectral changes and frequency modulations are components that influence the conspecific's succeeding vocal behavior. It may be concluded that these components, produced by the squirrel monkeys, are indeed essential for intraspecific communication. If this statement can be accepted, the method presented in this chapter may serve as a basis for establishing a "natural" categorization of the vocal repertoire of primates more reliable than one based on observer decision. Although we are only at the very beginning of our investigations, and despite the fact that only more or less arbitrarily chosen structural components have been presented here, it has become evident that the squirrel monkey's social behavior is influenced by acoustic structures other than those determined in earlier investigations.

Two major issues have not been taken into consideration as yet: (1) Are the structural components found to be functionally different exactly those the monkeys use? Were are the (categorical) boundaries? Are they mixed with components irrelevant for information transmission? How do (essential or inessential) components structurally and functionally interfere with each other? How much are they affected by peculiarities of the communicators? More generally, to what extent does the approach offered in this chapter help to analyze the interrelationships between structures and functions of acoustic patterns? (2) If this approach leads to a list of functional units clearly defined by structural properties, will it also be helpful in exploring the communicative implications of the units found, that is, in dealing with the functions themselves?

6.1 Interrelationships Between Structural and Functional Qualities

Instead of relying on observer decisions and *attributing* functional properties to structural patterns or vice versa, we begin by *defining* structural patterns. These patterns are not chosen entirely at random, but primarily those components are selected for testing their influence on the conspecific which seem to be of communicative relevance (e.g. Maurus et al. 1984; Barclay et al. 1987; Streit et al., in prep.). If we find a component to have a function, we may or may not have hit upon the structural category exactly within its boundaries; it can be smaller or larger, overlapping on either border, or it can even be an entirely different component more or less connected with the one we believe to have analyzed. How can our approach solve problems of this kind? An amplitude leap, for instance, was defined as being a sudden, at least fivefold fall or rise in amplitude modulation. This definition was derived from the characteristics of amplitude leaps as produced by the monkeys. While this definition may be of functional significance, it is quite possible that, to a monkey, the categorical boundaries are four- or sixfold falls or rises. By changing the definition and retesting each of the newly defined components, we are able to describe the structure more precisely with respect to function: the more significant the influence on the conspecific's behavior, the better our structural approximation of the monkeys' "natural" categories.

If the structure we have analyzed is associated with another communicative component, these two components may overlap at either (categorical) boundary. They may have altogether different functions, one may exclude the other, affect the other or cause a functional subdivision of the other. In all these cases functionally interfering components may be analyzed by separating them with regard to structural patterns presumably causing the interference and testing each of these patterns for functional differences (see this work: changes of power spectrum, and Streit et al., in prep.). There is the possibility, of course, that we attributed a functional influence to the wrong component part, that this influence is in fact exerted by another one always produced together with the one we selected. This would be a negligible error, however, for it would not affect the accuracy of the functional categorization and the usefulness of the structural description.

6.2 Individual Peculiarities of Communicators and Experimental Conditions

The aim of our investigation is to develop methods for finding the "natural" categories in vocal behavior of primates. It is not our primary objective to describe in detail the vocal behavior of individuals, groups, populations, subspecies or species under exactly defined circumstances or experimental conditions. We have no doubt, however, that such details can affect our results significantly. Individual peculiarities can be variations in sound production and perception due to anatomical or physiological conditions which can be modified, for instance, by confinement, experimental circumstances, social situations, temperature, humidity, etc. At the present state of our investigations, the variations caused by individual peculiarities can be separated from the modifications of vocal communication processes caused by environmental and social circumstances only theoretically. However, such theoretical separation and that of *what* and *how* components (see p. 75) may lead to a discussion helping to make the degree of influence these issues have on our findings clearer to us.

Each experimental situation is handicapped by the absence of natural circumstances. There may be calls the monkeys utter more often, much less or not at all in experiments as compared to field conditions. Calls not produced in experiments, of course, will not yield results and will make the investigation of the species' repertoire incomplete. This situation may be improved by modifying the experimental setup. Our method of measuring the influence that a call component has on the conspecific is highly independent of experimental conditions, provided the monkeys "dialogue" and our recordings as well as analyses are performed correctly. There is hardly any indication, if monkeys are together with conspecifics only in captivity, that they learn to produce and receive new call components and use them to influence each other. We can be quite sure that experimental conditions will not cause a change in the communication rules.

Difficulties may arise when monkeys are used that come from different areas, tribes, populations or subspecies. Due to the limited information received from dealers, this seems to be an unavoidable problem. If the monkeys participating in the experiments do not "understand" each other, of course there will be no proof of influence. The results we obtain when they "understand" each other are treated similarly to those obtained from investigating individual peculiarities (see below).

The collection of data from numerous experiments with different animals is necessary to find call components valid for all monkeys (*what* components). Pooling the data will hardly present results not actually existing; what it may do is blur results.

Data coming from monkeys not using the *what* components will not contribute to our results and consequently will have no contradictory effect. If results indicate that a monkey does not use the component and the range of the structural pattern was chosen correctly (i.e. the range limited by components' structural boundaries), then the monkey does not use this *what* component at all or not when together with this particular conspecific. If the range of the component was taken too narrowly, however, this "non-use" may be a delusion caused by the fact that the monkey produced the *what* component within the range we have mistakenly excluded when defining the component. Such a mistake can be eliminated by gradually varying the range and reexamining the influence on the conspecific. As stated before, by this procedure it is hoped to arrive at a closer correspondence of the component defined by us with the monkeys' "natural" category. According to our theoretical separation, this overall range would be the *what* component, while the range of each individual would indicate *how* the individual performs the component. The *how* structure can be identical with the *what* structure for all animals. It can be different for each animal and stable. It can also vary depending on experimental conditions, on the conspecific's identity or on the behavioral context.

Dependence on experimental circumstances can only be investigated by varying these circumstances. We keep these circumstances as constant as possible in order to gain data for analyzing more interesting aspects, such as the effect the conspecific's identity or behavior has on the interrelations between structures and functions. We have good reason to assume that squirrel monkeys are able to recognize a vocalizer that they can neither see nor smell (Maurus et al. 1987). Consequently, there is information about the vocalizer in the call. Is it in the component tested (i.e. is it an individual peculiarity of the component structure tested?) or is it in another call component? A hypothetical example may aid in seeking an answer to this question.

We know that the existence of exactly one amplitude leap is an essential component separating the function of that call from the function of a call with more than one amplitude leap. An amplitude leap can be located at the beginning, around the center or toward the end of a call. Now, if monkey A predominantly produces calls with a leap at the beginning and monkey B produces calls with a leap toward the end, we can define a leap at the beginning of a call to be a component different from a leap toward the end of a call. A comparison of these two components with the data pooled will either reveal significant differences or none at all.

In the first case, we can conclude that both components are *what* structures because they differ in their influence on conspecific behavior regardless of who the sender is and who the recipient. The two components are structural and functional subdivisions of the component from which they originated. In the second case, they may either be irrelevant or may transmit information on who the partner is. This *how* performance of the original *what* component may possibly contain the sender's identity, some other information or none at all. However, use of this component will be restricted to this particular animal. We may speculate that it transmits the sender's identity when it is stable and that it is of more significance when it changes, depending on the social situation, the conspecific's identity, etc. If monkey A predominantly produces leaps at the beginning of a call when "dialoguing" with animal B and toward the end of a call when "dialoguing" with animal C, and if leaps at the beginning of calls are functionally different from leaps toward the end of calls, it would mean that, depending on who the conspecific is, the monkey communicates with different *what* components (different topics?). In case leaps at the beginning and leaps at the end of calls are functionally the same, the monkeys would communicate with identical *what* components, but *how* they communicate *what* depends on the receiver's identity (*how* to communicate *what* topic is determined by the conspecific, too).

When and how our approach can be helpful may have been illustrated by the example described above. Comparing results gained from one individual and from pairs with those gained from pooled data is only one of the many differentiation possibilities. Another, more sophisticated approach is based on the refinement of the scale of measurements. Instead of roughly differentiating between x and z calls, each pair of components proven to differ in function can be used for measuring the influence on the conspecific. If we know, for instance, that x_i is functionally different from x_j, and we want to test x_r versus x_s, we may roughly test $\frac{xx_rx}{xx_rz}$ versus $\frac{xx_sx}{xx_sz}$ or, for more refined measuring, $\frac{xx_rx_i}{xx_rx_j}$ versus $\frac{xx_sx_i}{xx_sx_j}$, etc. The more specified the questions are and the more refined the scale of measuring is, the more data we need in order to obtain samples in sufficient quantities to prove the reliability of our results.

6.3 Functional Interrelationships

Analysis of the interrelations between structures and functions can help to clarify the mechanisms underlying primate vocal communication, but it is only part of what an ethogram should be. After defining the structural patterns comprising functional categories, the interrelationship between the functional categories and their ethological labeling should be determined. But while our measurements tell us which structures differ in function and where the differences are, they do not say anything about the types of differences. For this reason, we have developed methods which have already proven to be successful (e.g. Pruscha and Maurus 1976; Maurus and Ploog 1984). Although these methods have been applied to visually recognizable signals, they can also be taken into consideration for acoustic signals. They are based on the comparison of categorized functions regarding succession in communication processes, the participating animals, social relationships, etc. Additionally, since all experiments were filmed, visually recognizable behavior, signal displays, as well as location and interindividual distance (Maurus and Ploog 1969), can be evaluated and evaluations combined with the results gained from acoustic analyses. There is no doubt that visual and acoustic communications are not entirely independent from each other. Interdependences are likely to exist and may be considered in interpreting the functional categories from an ethological point of view.

As explained in the introduction (p. 70), structural analysis begins by splitting the repertoire, and the finer it is split, the smaller is the danger of bias. Our theoretical differentiations (such as *what* vs *how*, structural vs functional analysis, interactions between structures and functions vs those between functions), of course, allow us to use knowledge obtained at one step again for another, more refined analysis yielding more accurate results.

6.4 Playback Experiments

Playback experiments thus far were successful when the animal's responses could be measured by other means than vocal utterances with unknown functions. When measuring the communicative influence conspecific vocalizations interrupting intraindividual transitions have, the use of playback vocalizations could be advantageous for they make the investigator independent of the conspecific's motivation to vocalize and, in addition, the vocalizations are reduced to only one or a few call types. We can even test the communicative significance of defined call components by excluding them from a call, adding them to a call or substituting them by components of any kind, thus achieving a manipulated, or completely synthesized acoustic structure. A disadvantage may be that the calls selected for playback (as long as we do not have an on-line analysis of the vocalizations that the experimental animal utters) may be inappropriate for the calls that the test monkey has uttered. Our results from experiments with conspecifics out of visual and olfactory contact (conditions closely resembling those of playback experiments) indicate that the readiness to "dialogue" depends very much on the partner's vocalization (see Fig. 3, Maurus et al. 1987). Consequently, it seems reasonable to use only vocalizations as playbacks which stem from conspecifics known to stimulate the test animal's response motivation. It should be investigated how much this practice affects restrictions in the use of manipulated or synthesized signals and to what extent the advantages of playback experiments exceed those of experiments where animals are joined. Whatever the outcome of such investigation may be, it seems worth the effort.

Acknowledgments. The authors wish to thank Mrs. E. Roesner for her assistance in translating the text.

References

Barclay D, Streit K-M, Maurus M, Steinleitner M, Haeglsperger R (1987) Linguistic strategies to detect frequency modulation patterns relevant to primate vocal communication. Lang Commun 7:255–265

Cheney DL, Seyfarth RM (1982) How vervet monkeys perceive their grunts: field playback experiments. Anim Behav 30:739–751

Green S (1975) Variation of vocal pattern with social situation in the Japanese monkey *(Macaca fuscata)*: a field study. Primate Behav 4:1–102

Hockett CF (1960) Logical considerations in the study of animal communication. In: Lanyon WE, Tavolga WN (eds) Animal sounds and communications. A J B S, vol 7. Washington D.C., pp 392–430

Masataka N (1983) Categorical responses to natural and synthesized alarm calls in Goeldi's monkeys *(Callimico goeldii)*. Primates 24(1):40–51

Maurus M (1974) Problems in the analysis of agonistic behavior in the squirrel monkey: an attempt at a solution by means of telestimulation technique, telemetry, and statistical methods. In: Holloway RL (ed) Primate aggression, territoriality, and xenophobia. A comparative perspective. Academic Press, New York San Francisco London, pp 331–359

Maurus M, Ploog D (1969) Motor and vocal interactions in groups of squirrel monkeys, elicited by remote-controlled electrical brain stimulation. Recent Adv Primatol 3:59–63

Maurus M, Ploog D (1984) Categorization of social signals as derived from quantitative analyses of communication processes. In: Harré R, Reynolds V (eds) The meaning of primate signals. Cambridge Univ Press, Cambridge, pp 226–241

Maurus M, Szabolcs J (1971) Kleinstsender für die Übertragung von Affenlauten. Naturwissenschaften 5:273–274

Maurus M, Pruscha H, Wiesner E, Geissler B (1979) Categorization of behavioural repertoire with respect to communicative meaning of social signals. Z Tierpsychol 51:48–57

Maurus M, Streit KM, Geissler B, Barclay D, Wiesner E, Kuehlmorgen B (1984) Categorical differentiation in amplitude changes of squirrel monkey calls. Lang Commun 4 (3):195–208

Maurus M, Kuehlmorgen B, Wiesner E, Barclay D, Streit K-M (1985) "Dialogues" between squirrel monkeys. Lang Commun 5 (3):185–191

Maurus M, Streit K-M, Barclay D, Wiesner E, Kuehlmorgen B (1986) Interrelations between structure and function in the vocal repertoire of Saimiri. Asking the monkeys themselves where to split and where to lump. Europ Arch Psychiat Neurol Sci 236:35–39

Maurus M, Kuehlmorgen B, Wiesner E, Barclay D, Streit K-M (1987) Vocal interactions between squirrel monkeys out of visual, tactile and olfactory contact. Lang Commun 7:39–45

Müller-Preuss P, Maurus M (1985) Coding of call components essential for intraspecific communication through auditory neurons in the squirrel monkey. Naturwissenschaften 72:437

Newman JD (1985) Squirrel monkey communication. In: Rosenblum LA, Coe CL (eds) Handbook of squirrel monkey research. Plenum, New York London

Pruscha H, Maurus M (1976) The communicative function of some agonistic behavior patterns in squirrel monkeys: the relevance of social context. Behav Ecol Sociobiol 1:185–214

Reynolds V (1976) The origins of a behavioural vocabulary: the case of the rhesus monkey. J Theory Soc Behav 6:105–142

Seyfarth RM, Cheney DL (1984) The acoustic features of vervet monkey grunts. J Acoust Soc Am 75 (5):1623–1628

Sweeting PM, Baken RJ (1982) Voice onset time in a normal-aged population. J Speech Hear Res 25:129–134

Zlatin MA (1974) Voicing contrast: perceptual and productive voice onset time characteristics of adults. J Acoust Soc Am 56 (3):981–994

Serial Calling as a Mediator of Interaction Processes: Crying in Primates

DIETMAR TODT[1]

The crying behaviour of primates is a conspicuous marker, and also an interesting mediator of interaction processes among group members, primarily between an infant and his caregiver. Various studies have verified that crying signals a distress state, and thereby calls for support through an ally (citations in: Lester and Boukydis 1985; Bisping 1986). Additionally it has been shown that crying in non-human primates can be addressed not only to a social ally, but at the same time may be directed towards another conspecific, which e.g. has harassed the signaller. Here, crying may also *repel* a distinct individual (Green 1975; Gouzoules et al. 1984; Todt 1986; Hammerschmidt et al. 1987).

Current hypotheses on the communicative significance of crying agree upon such double function; but they also reflect the various conceptual views fostered by different investigators.

1 Aspects Emphasized by Psychologists

The majority of studies on human infants were designed to elucidate two basic aspects of crying: (1) the particular "meaning" of this ontogenetically early display (Brazelton 1962; Bosma et al. 1965; Truby and Lind 1965; Wasz-Höckert et al. 1968; P. Wolff 1969; Emde et al. 1976; Michelsson et al. 1977; Golub and Corwin 1982; Lounsbury and Bates 1982; Grossmann 1985; Lester 1985), or (2) its impact on the development of infant/caregiver relationships (Bell and Ainsworth 1972; Hassenstein 1973; Murry et al. 1975; Frodi et al. 1978; Lester and Zeskind 1979; Zeskind 1980; Frodi 1981; Sagi 1981; Brennan and Kirkland 1982; Brüser 1985; Boukydis 1985; Donovan and Leavitt 1985; Hammerschmidt et al. 1987).

A rather general concept pointed to the acoustic conspicuousness of crying, and predicted that its noise properties could bring nearby listeners into aversive distress. So, listeners would have two choices: Either to leave, or to soothe the crying individual (Ostwald 1963, 1972).

Proceeding from this *"noise hypothesis"* psychologists have argued that crying should lead to a mutual conditioning of interactors. In particular, the care giving of adults gets positively reinforced, if it terminates the crying of an infant. To explain such types of learning an "egoistic motivation" has been postulated, which, however,

1 Institut für Verhaltensbiologie, Fachbereich Biologie, Freie Universität Berlin, Haderslebener Straße 9, 1000 Berlin 41, FRG

Todt/Goedeking/Symmes (Eds.)
Primate Vocal Communication

was criticized as being insufficient for an adequate interpretation of the various kinds of parental care (Hoffmann 1975; Murray 1979). Therefore, this concept was supplemented by the *"empathy hypothesis"* (Murray 1985). It states that the care-giving motivation of adults stems from their sympathetic distress which arises due to the distress signalled by an infant. Support for this hypothesis came from results showing that a mother's anticipation effects her responses; thus, she responds more strongly when she was told beforehand that her infant is in pain than when she was told he is hungry (Zeskind 1980; Lester and Zeskind 1982; Lester 1985; Murray 1985).

An interpretation of crying which, in particular, emphasized the interactive role of the infant has been derived from the *"developmental attachment theory"*. According to Bowlby (1969) this display behaviour cannot be understood without considering the infants' perspective. As stated by Ainsworth, crying behaviour is a mediator for the development of social bonds (Ainsworth 1973; Ainsworth et al. 1974). It allows the infant to acquire interaction experience at an early stage, and to support orientation even within intricate social contexts (*attachment hypothesis*).

Crying behaviour is a universal display of human infants, but regional, as well as individual differences in its occurrence are evident. Regional differences have been traced back to cultural diversities which have an impact on mother-infant relationships and qualities of the care-giving context (Ainsworth et al. 1974; DeVore and Konner 1974; Boukydis 1985). Individual differences were found to depend on long-term variables such as temperament, genetic or developmental deficiencies or illness (Lind 1965; P. Wolff 1969; Frodi 1985; Lester 1985; Liebermann 1985; Murray 1985; Ostwald and Murray 1985). Besides theses aspects, also the intra-individual variability of crying has been studied, e.g. in terms of cues indicating internal components, such as affect or arousal (Lester 1985).

2 Aspects Emphasized by Biologists

Vocalizations of infants uttered when these are separated from the mother (lost calls, isolation calls, separation calls or various kinds of cry calls) share fundamental similarities among different primate species, including human beings (Gautier and Gautier-Hion 1977; Oppenheimer 1977; Liebermann 1985; Newman 1985). Studies concerned Prosimians, such as the genera *Galago* and *Lemur* (Andrew 1963), or Platyrrhini, such as *Saimiri sciureus* (Lieblich et al. 1980; Newman and MacLean 1982; Newman and Symmes 1982), *Alouatta palliata* (J.D. Baldwin and J.I. Baldwin 1976), *Ateles fusciceps* (Eisenberg 1976), *Callithrix jacchus* (Pook 1977), *Cebuella pygmaea* (Pola and Snowdon 1975), *Oedipomidas spixi* (Epple 1968), *Saguinus fuscicollis* (Hodun et al. 1981) and Catarrhini, such as *Cercopithecus aethiops* (Struhsaker 1967), *Macaca mulatta* (Rowell and Hinde 1962; Gouzoules et al. 1984), *Macaca arctoides* (Chevalier-Skolnikoff 1974), *Macaca nemestrina* (Simons and Bielert 1973), *Macaca fuscata* (Green 1975), *Macaca silenus* (Green 1981), *Macaca sylvanus* (Todt et al. 1985; A. Wolff and Todt 1985; Hammerschmidt et al. 1987), and also the Hominoidea *Gorilla gorilla* (Fossey 1979) and *Pan troglodytes* (van Lawick-Goodall 1969). Since infants' signals in the above context share the basic pattern of a gradually frequency-modulated tonality, Newman (1985) has concluded that the mechanisms controlling infant cry patterns have had a rather conservative evolutionary history.

Primate infants are extremely attractive for other group members (Kummer 1967; Deag and Crook 1971; Hrdy 1976; Suomi 1977; Taub 1980; Haut and Todt 1984; Nicolson 1987; Whitten 1987). Therefore it is unlikely that an infant would ever get isolated from his group. Separation from his mother, on the other hand, may occur frequently, namely as a result of the various kinds of allomothering. However, as the strong interest of group members towards infants often occurs independently from adequate infant handling and care-giving ability, I have hypothesized that crying behaviours may have evolved as a consequence of allomothering.

The *"allomothering hypothesis"* is based on the assumption that an infant can solve an interactional problem with an incompetent allomother only by display behaviours. Thus, there are good reasons for the development of signal structures which allow one to repell a group member harassing the infant and, at the same time, to call for help from a social ally, in particular the mother.

Studies on crying in Barbary macaques *(Macaca sylvanus)* support the allomothering hypothesis. In this species, infants are preferred targets of various behaviours of adults and subadults, which often harass them. For example, infants may have problems returning to their mothers, and crying occurs quite frequently until the age of about 18 months (Todt et al. 1985; A. Wolff and Todt 1985). In this context, crying vocalizations are clearly linked with interactional distress. So, crying may be a compromising interactional strategy: the benefit obviously counterbalances risks eventually arising from its conspicuousness also for potential predators.

Hypotheses on evolutionary perspectives emphasized that crying functions as a call for contact and support, but did not consider its repelling functions mentioned above. There is evidence that special signal features, such as atonal and noisy patterns, mediate a repelling of harassers. Such features are widespread in cry calls of primates, and occur especially often at an evolutionary, early stage (Andrew 1963). These features are not optimal for the contact function of crying, because they are strongly affected by signal attenuation, and consequently do not penetrate very far. But, they may generate aversive distress in nearby recipients; eventually in the same manner as postulated for human beings by Ostwald (1972).

Referring to process features typical for crying in primates, such as wide frequency range, high sound pressure level and extended duration of vocalizations, I propose to supplement Ostwald's hypothesis by a crucial biological aspect: There are sound biological reasons to explain the aversive distress which crying may cause in listeners by the fact that the *auditory channel* of any proximate conspecific becomes *occupied* and, consequently, its ability to auditorily check or monitor the environment is impeded. If the cry is used against a harasser (or an incompetent allomother), this constraint should be particularly risky, since this individual – due to its proximity to the caller – can easily get localized by other group members, and e.g. get displaced by the caller's allies. Nevertheless, we have to ask whether cry features may similarly affect nearby allies, and so also raise the probability of their responses.

3 Structural Aspects

Acoustically, crying behaviour can be characterized as a process which is constituted of a series of expiratory signal patterns termed *"cries"* (synonymous in non-human primates: "screams"). In human infants normal cries (P. Wolff 1969: proper cries) have a mean duration of 0.5–1.5 s, and are mutually separated by intervals normally shorter than the preceding cry. These inter-cry intervals are either silent pauses or include a short and higher pitched inspiratory sound (mean duration: 0.1–0.2 s). In non-human primates such inspiratory sounds are less obvious, and both cries and inter-cry intervals may be shorter than in human beings.

Due to the conspicuousness of the expiratory vocalizations, most studies have focussed on these signal patterns only. For crying in human infants studies distinguished between "basic cries", also termed "phonations", and "intensified cries", also termed "paraphonations" such as dysphonations and hyperphonations (Lind 1965; Brüser 1985; Lester 1985; Bisping 1986; Fischer 1988).

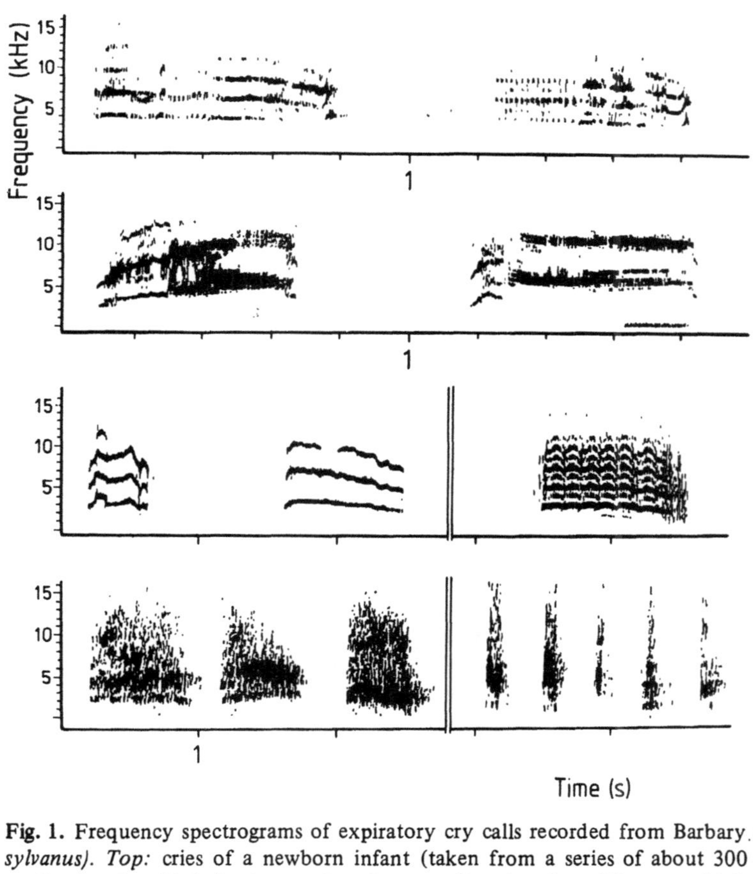

Fig. 1. Frequency spectrograms of expiratory cry calls recorded from Barbary macaques *(Macaca sylvanus)*. *Top:* cries of a newborn infant (taken from a series of about 300 calls vocalized immediately after birth in the monkey forest at Roc Amadour (distance of infant to microphone: approximately 3 m). *Second row:* cry calls vocalized by an infant a few weeks old. For comparison, examples of more elaborated cry call types vocalized by older infants (about 1 year of age) are given below. *Third row:* tonal (*left*) and undulated (*right*) cry types. *Fourth row:* noisy (*left*) and pulsed (*right*) cry types

For a finer discrimination among cries typological categorization methods provided a well-proved means, even if clear-cut boundaries between different types were difficult to ascertain (Hassenstein 1955; Todt 1968, 1986; Green 1975; J. Wolffgramm and Thimm 1976; Jürgens 1979; Hultsch 1980; see also Maurus, this Vol.). In our studies we have distinguished among the four basic types: tonal cries, undulated cries, noisy cries and pulsed cries. Often, such basic features were combined, thus we distinguished the following "mixed types": tonal-pulsed cries, tonal-undulated cries, tonal-noisy cries, tonal-noisy-tonal cries (Figs. 1, 2, 3).

On the basis of signal categorization, the evaluation of crying processes recorded from different primate species (including human beings) revealed a widespread rule: subsequent cries tend to belong to the same cry type (Jürgens 1979; Gouzoules et al. 1984; A. Wolff 1985; Hammerschmidt et al. 1987). Consequently, crying behaviours, or at least bouts of successive cries, can often be regarded as homo-type call series. Exceptions may occur during an initial process phase (P. Wolff 1969), or in the case of extremely prolonged crying. So, when infants cry longer than 1 min, they may switch to another cry type; thus, performing a hetero-type process which consists of two or more homo-type series. A further possibility is illustrated in Figs. 1 and 2.

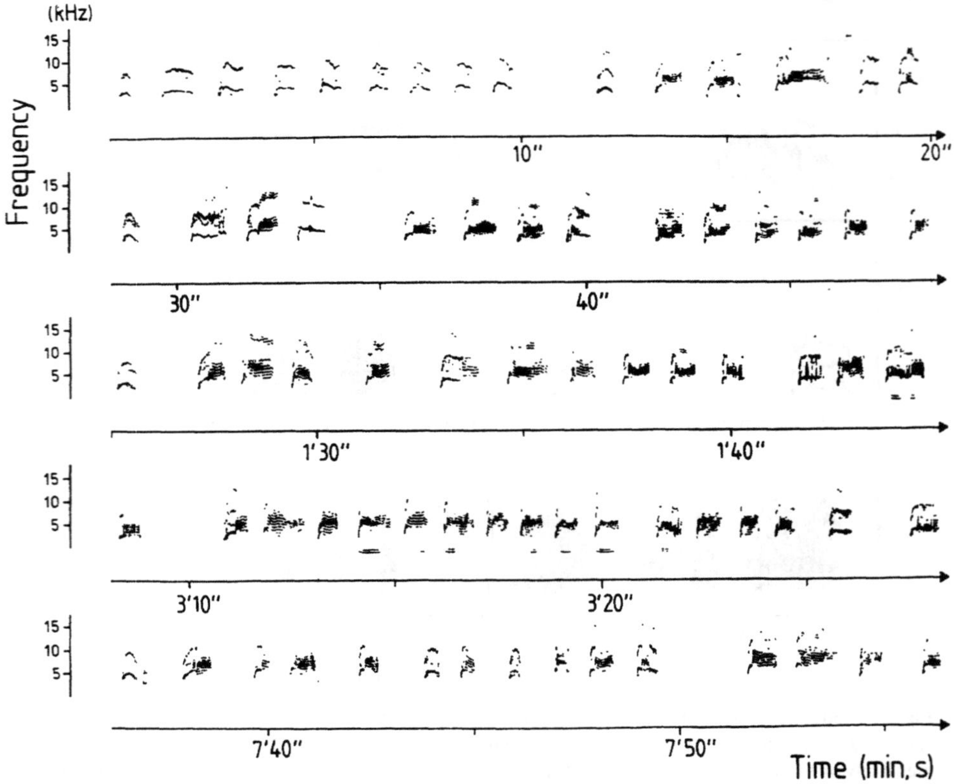

Fig. 2. Frequency spectrograms of a crying process recorded from a 1-year-old Barbary macaque infant. Here, crying is initiated by a number of tonal types. Later, cries exhibit a rather mingled characteristic (e.g. tonal-undulated or tonal-noisy)

It seems conclusive that the particular structural features of crying should have an impact on its functional properties. For instance, the silent intervals between successive cries are often remarkably shorter than those normally found between the various kinds of calls (or songs) which typically serve a mutual exchange of messages (Hultsch and Todt 1982; Wolffgramm and Todt 1982). Therefore, the communicative role of crying should be quite different from the role ascertained for the latter. For the same reasons, a coherent string of cries can be regarded as a *super-signal*, wherein both the *sequential* and *temporal organization* (syntax) of constituents may be communicatively significant, and so also effect the interactional role of crying (Todt and Hultsch 1980).

As studies on other forms of serial calling have documented, temporal variations of acoustic parameters can reflect interesting rules: for instance, signal parameters of serially uttered vocal units in birds can be altered predictably in timing (accelerandos, decelerandos) or volume (crescendos, decrescendos) (Thorpe 1972; Hultsch 1980, 1983). Alterations can gradually approach particular "Sollwert" ranges. There is evidence that besides absolute parameter values upon approaching the "Sollwert" also the gradients of parameter variations may serve particular social functions (Todt et al. 1981; Todt 1986). In primates, temporal dynamics which are socially significant have been reported

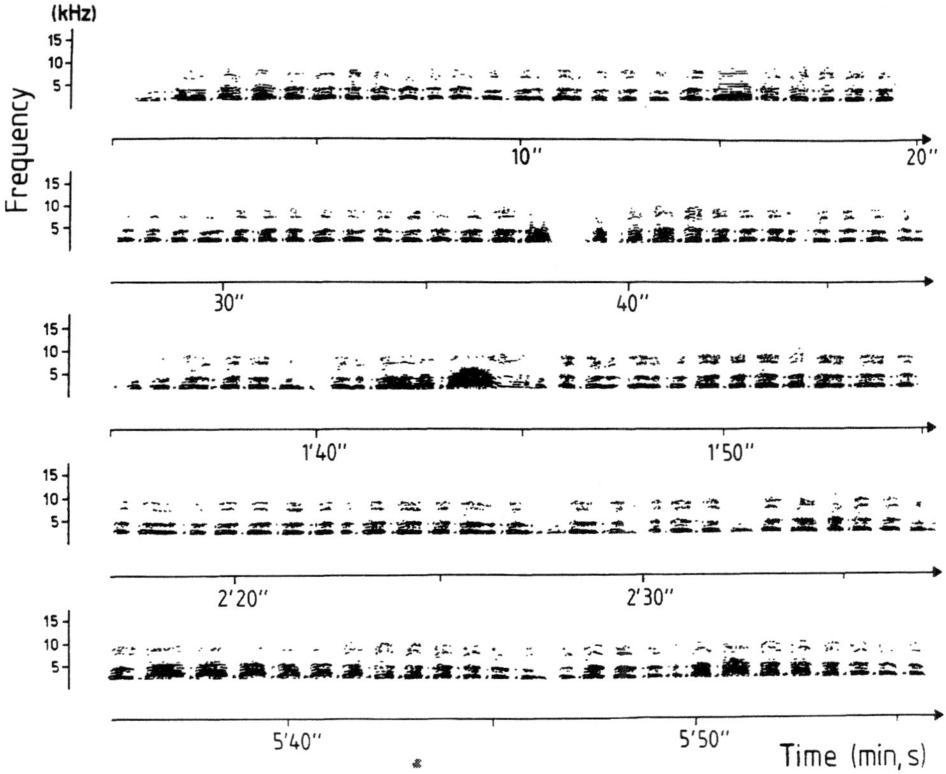

Fig. 3. Frequency spectrograms of a crying process recorded from a 2-week-old human infant. Here, inspiratory sounds were uttered between consecutive expiratory cry calls

for the serially structured mating calls of Barbary macaques (*Macaca sylvanus*, Todt and Pohl 1984). But hitherto, such findings have not been documented for the crying behaviours of primates. In the following section data are presented which examine this matter.

4 Dynamics of the Crying Processes

To test temporal dynamics occurring during a crying process, we conducted two kinds of approaches: (1) a fine-graded analysis of successive cries by comparing relationships among temporal micro-segments of the calls. This procedure supplemented the comparison of complete cries (see preceding section). (2) An analysis of selected crying parameters which particularly focussed on their temporal variations. To allow for a comparison among process constituents the majority of tests concentrated on homotype processes.

For the analysis of *temporal micro-segments* sound spectrograms of cry calls were cut into successions of temporal "slices" (micro-segments; duration either 100 or 250 ms, depending on the issue concerned) and each segment was categorized to one of the following five classes: tonal, undulated, noisy, pulsed or "rest" (e.g. mingled features). Classed segments were analyzed according to both their particular positions and their sequential succession within a cry.

The evaluation of data provided three results, which indicated trends of an increasing signal variability along the ongoing crying process (Figs. 4, 5):

1. Micro-segments located at the beginning of successive cries were more similar to each other than were segments positioned in the middle or near the end of these cries.
2. Similarity was highest among the equally positioned micro-segments of sequentially close cries, and decreased with the sequential distance of such cries.

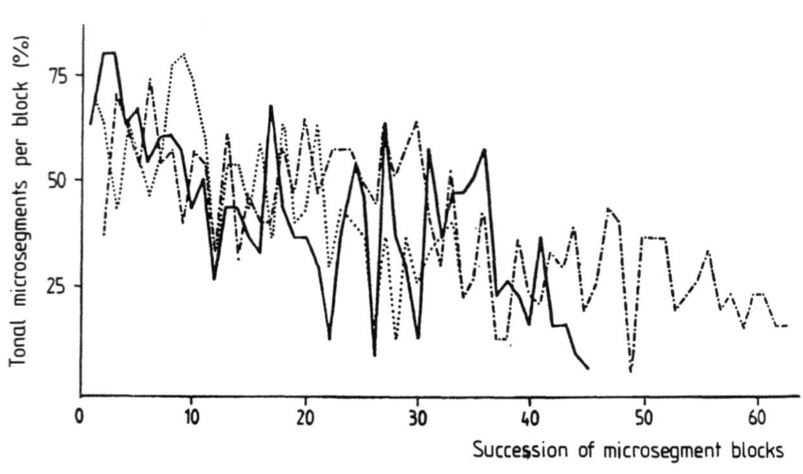

Fig. 4. Decrease of tonal microsegments during three different crying processes recorded from a human infant. The relative frequency of tonal microsegments (duration: 250 ms) in a block of 20 subsequent microsegments is plotted against the succession of these blocks

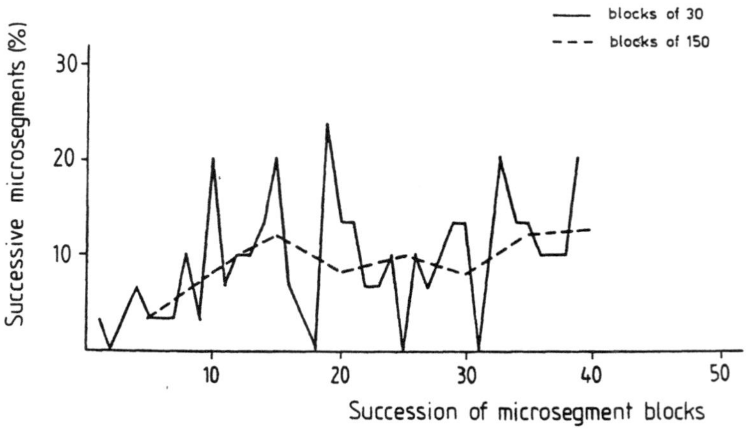

Fig. 5. Variability of acoustic features during a cry process. Here, the relative frequency of differently structured microsegments that (within either a block of 30 or 150 microsegments) initiated successive cry calls are plotted against the succession of blocks

3. Taking together the set of micro-segments which made up a complete cry, we found that successive cries located at the beginning of a process were more similar to each other than were successive cries located at the end of the same process.

Interestingly, the effect of a progressive variability increasing on two hierarchical levels (within and between cries) could be evidenced in crying behaviours of both human infants, and young Barbary macaques *(Macaca sylvanus)* as well (Hammerschmidt et al. 1987).

For the analysis of *temporal parameter variations*, values of the following variables were ascertained for each subsequent call of a crying process: (1) the fundamental frequency; (2) the frequency range; (3) the repetition interval (measured in ms from the onset of a cry to the onset of the next one); (4) the "acoustic density" (duration of a call over both the duration of this call plus the duration of the subsequent silent interval); (5) the sound pressure level. The analysis focussed on both the particular variation of the respective parameter and the temporal dynamics of these variations.

Dynamics were examined by techniques of time series analysis (Schlittgen and Streifberg 1987). Calculations of trends followed the "minimum square criterion" method.

Normally, the calculations revealed consistent parameter trends from the onset to the end of a crying process. Data showing trend examples are given in Fig. 6.

In addition to trends covering the whole process, two other kinds of parameter variations were found. These were "subtrends" which often reached intermediate maxima (or minima), and also instances of "extreme parameters increases" (EPIs) identified by immediate value increases which exceeded two quartiles of the particular distribution (Fig. 6).

The covariation of parameter dynamics was studied especially for parameters which contributed to the acoustic conspicuousness of cry processes. As shown in Table 1, the

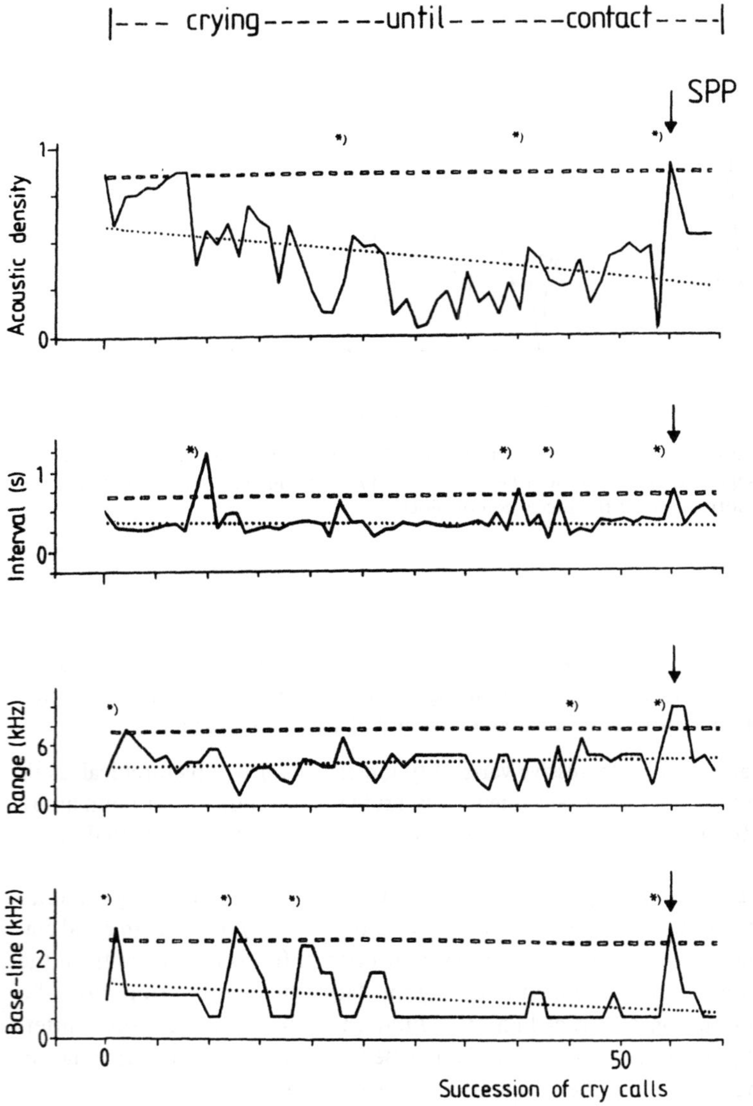

Fig. 6. Parameter variation in a crying process vocalized by a Barbary macaque infant. *Dashed double lines* give the upper 90% level of parameter changes. *Dotted lines* show the calculated parameter trends. *Asterisks* indicate positions of extreme parameter increases (EPIs). A synchronous parameter peaking (SPP) is expressed at the end of the process. This end coincided temporally with the establishment of body contact between mother and infant

variation lacks a positive correlation among the following parameters; peaking amplitude, acoustic density and frequency range.

Two aspects make this result particularly interesting. The exchangeability of parameters that take a maximum value raises the variability of the signalling process without similarly reducing its conspicuousness. In addition, it allows a rather economical manner of applying vocal parameters.

Table 1. Correlation between increases of acoustic parameters ascertained for a crying process recorded from a Barbary macaque infant (significance: $^{*}p = 0.01$; $^{**}p = 0.001$)

Parameter	Peaking amplitude	Acoustic density	Repetition interval	Call duration	Frequency range	Base-line	Upper frequency
Peaking amplitude	1	0.17	–0.03	–0.3**	–0.01	–0.13	–0.14
Acoustic density	0.17	1	0.45**	–0.22	0.09	–0.13*	–0.05
Repetition interval	–0.03	0.45**	1	0.58**	0.05	0.07	0.07
Call duration	–0.3**	–0.22	0.58**	1	0.02	0.25*	0.15
Frequency range	–0.01	0.09	0.05	0.02	1	0.09	0.86**
Baseline	–0.13	0.23*	0.07	0.25*	0.09	1	0.58**
Upper frequency	–0.14	–0.05	0.07	0.15	0.86**	0.58**	1

A third aspect results from occurrences of synchronous increases of several parameters which leads to instances of "synchronous parameter peaking" (SPPs; Fig. 6). Presumably, the kind of correlation among parameter expressions given in Table 1 provides a "baseline" for particular contrast phenomena reached through the SPPs. I assume that these contrast phenomena may serve to raise and focus a recipient's attention immediately. Experiments conducted to test this hypothesis will be reported in the following section.

A well-proved methodological step in a formal analysis of behavioral data is the examination of relationships among verified effects (test for independent occurrence: Todt and Wolffgramm 1975; Todt and Hultsch 1980). From such tests it can be concluded that both the EPIs and the SPPs were a sort of behavioural dynamic which occurred independently of a particular parameter trend. Additionally, it has to be emphasized that the temporal dynamics found in the analyzed processes did not correlate with changes in contextual variables. Therefore, the revealed dynamics have to be interpreted as a reflection of a signaller's internal state variables.

5 Cues from Responses of Recipients

There are numerous studies on crying which tried to ascertain signal messages by examining recipients' responses (citations in: Donovan and Leavitt 1985; Bisping 1986; Fischer 1988). However, most of these studies used a single cry call as a basic stimulus, and only a few of them considered the temporal dynamics or other process properties of crying. For the measurement of responses, investigators favorized either physiological methods (e.g. analysis of heart rates; Bryan and Newman 1988), or psychological methods (e.g. evaluation of questionaries; Zeskind and Marshall 1988), or behavioural methods (e.g. vocal responses; Brüser 1985). Based on a behavioural *online approach* which will briefly be summarized in the following, the current research of our laboratory applies a test setting that allows a synchronous examination of 10–20 probands per session (Todt and Wiedenmann, unpublished).

Human adults (11 female and 19 male students) without own children were exposed to cry call series (duration: 1–5 min) by playback through a one-ear headphone. Recordings had been collected from human infants (age: 1–3 weeks) who were crying, before getting their food, and while their mothers were inside their rooms. For control, some of the probands were exposed also to synthesized cry calls and, additionally, asked to complete questionnaires. To prepare for the testing, probands were placed in a chair and given three orders: (1) To imagine taking over the job of a baby-sitter. (2) To turn their head to the side of the one-ear headphone (normally fitted on the right-hand side) as soon as they had the impression that "their" crying infant should be cared for and nursed ("checking response"). (3) To slightly lift one hand when they tended to judge the crying as "unpleasant" ("distress response").

Experimental data were collected by means of a two channel tape recorder which stored the crying process played back to the proband (channel I), and simultaneously recorded the proband's responses (onset and offset of "head turning" and "hand lifting", channel II). Tapes were analyzed according to relationships between response rates or temporal positions of responses and features of the concerned crying processes. Evaluation of data provided evidence for an interesting difference between rates and positions of responses.

Responses of probands (which, such as the "head-turning" towards the cry-sound source, signalled that they intended to check an infant's condition), were positively correlated to the following parameter configurations:

1. Suddenly *prolonged duration* of expiratory cries, which often was linked with a raised acoustic density.
2. Sudden *change* of *inspiratory* sounds resulting from either a drop or a modified position of these sounds.
3. Suddenly *accelerating rhythm* resulting from a shortening of repetition intervals of cry calls.
4. Sudden *increase* of the *sound pressure level* (peaking amplitude) of expiratory cries.

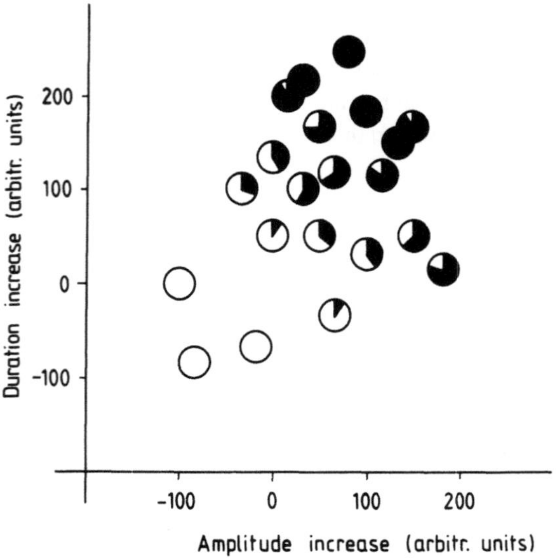

Fig. 7. Dependency of the probands' (N = 18) checking responses (here: head turning) on increases of two parameters (duration and peaking amplitude) of the expiratory calls. Increases are given in % of the difference between the parameter value of a particular call and the mean parameter value of two preceding calls. *Circles* refer to instances of observed differences. *Filled portions of circles* represent the relative amount of revealed responses. As shown, this amount was especially high when both parameters increased synchronously; and no responses were found when a decrease of one parameter was *not* "counterbalanced" by an increase of the other one (see *vacant circles*, given here for comparison)

5. Sudden *increase* of the *frequency range* of cries being linked with a *raised fundamental frequency*, and a considerable part of a cry reaching far above 6 kHz.
6. Occurrence of *noisy* cry types, or at least cries with bands being irregularly interrupted.

The predictability of these "checking" responses was particularly good for those process positions at which several of the parameter configurations given above (1–6) occurred simultaneously (e.g. so forming an SPP; Fig. 7).

In contrast, both a phasic and a tonic characteristic were ascertained for the "hand-lifting" responses of probands, (which, such as the hand lifting, signalled their unpleasant feelings; Fig. 8). As phasic responses, "hand lifting" was clearly related to a joint occurrence of parameter configurations 5 and 6 yielding cry patterns of a marked "shrillness". When "hand lifting" occurred at other positions of the crying process, these responses revealed a tonic characteristic: the probability of expressing distress increased parallel to an increase in "head-turning" rates (five or more per min). And, once elicited, the distress response "hand lifting" lasted for some time, often until the end of the experiment.

Further analyses showed that experimentally induced distress of our probands covaried with a specific aftereffect. This was ascertained as a raised probability (or, a lowered threshold) for distress responses in such probands who, in earlier trials, had been exposed to a series of shrill cry features imbedded in a synchronous increasing or peaking of the crying's acoustic density and amplitude (EPI, SPP). The raised probabilities could not be explained as a mere consequence of a short-term sensitization, but turned out to be a temporally persistent phenomenon.

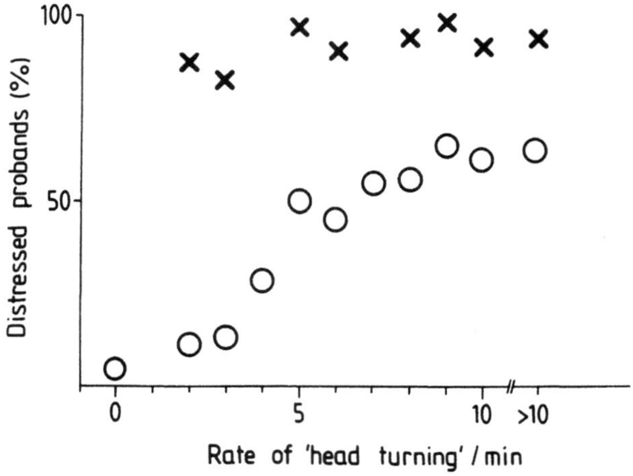

Fig. 8. Dependency of distress responses (measured by the probands' "hand lifting") on features of crying processes (*crosses:* crying *with* shrill features; *circles:* crying *without* shrill features). The number (%) of probands (N: 7) who signalled a "hand lifting" is plotted against the rate of their "checking" responses (here: head turning towards the cry-sound source). As shown, crying processes including shrill cry features elicited distress responses rather independently of the rates of "head-turning" (see text)

Table 2. Conditioned distress responses revealed by analyzing the time budgets of distress responses (here: "hand lifting") during a sequence of four experiments [pre test, test, post-test (1), (2)]

Probands	Amount of time (%) spent performing a distress response[a]					
	Pre-test mean (SD)		Test mean (SD)	Post-test (1) mean (SD)		Post-test (2) mean (SD)	
Test group (N = 9)	22 (7.2)	ns	37 (6.8)	31 (5.4)	+	14 (4.9)	+
Control group (N = 5)	19 (8.0)		21 (7.4)	22 (8.8)		6 (3.3)	

[a] *Test:* Auditory exposure of probands (test group) to a standardized record of crying (3 min) which contained a number of shrill cry features. *Pre-test* and *post-test (1):* Exposure to the same record, but after it had passed a filter (extinction of cry features above 6 kHz), thus the shrillness of cries was extremely reduced here. *Control:* Exposure of other probands to the filtered crying record in all three trials. *Post-test (2):* Exposure of both test group and control group to another crying record which did not contain shrill features. Time intervals between pre-tests, tests and post-tests (1), (2): 1 week. Significance: + p: 0.01; ns: not significant (Kolmogorov-Smirnov test).

Since probands, who had been exposed to SPPs including shrill cry patterns in earlier tests, were found to respond by indicating distress in later tests, even when hearing SPPs *not* including shrill patterns, we concluded: The latter responses can be interpreted as a kind of "conditioned distress responses", in which unpleasant feelings of the probands may be a consequence of an SPP-caused anticipation of shrill features (Table 2).

The evaluation of questionnaire ratings revealed the highest conformity when the probands could vote either for a "helplessness message" or a "protest message" expressed by a crying infant. "Helplessness" was attributed to a crying that could be characterized by the following acoustic features: (1) inspiration sounds were longer in duration and louder than normal, (2) expiration sounds were long in duration, tonal in structure and often showed a low-pitched frequency range. A "protest" was decoded from crying processes which, in contrast, (1) did not contain conspicuous inspiratory sounds, but showed both (2) atonal expiratory sounds and (3) frequent EPIs or SPPs that could induce a behavioural distress response. These findings indicated that the inspiratory sounds, in particular, may have a communicative function in crying behaviours, which had not been considered hitherto (Fig. 9).

The finding that raising the frequency spectrum of cries may result in an increase of aversiveness in adult listeners (or, in unpleasant feelings) is in good accordance with results gained by an experimental design which Bisping (1986) has called "operant titration". It allowed one to ascertain thresholds of aversion by the following procedure: Probands exposed to acoustic stimuli can operate a switchboard to reduce the sound pressure level if stimuli become aversive. The relationship between acoustic parameters of stimuli and switchboard operation of the probands revealed a striking role of the fundamental frequency in a cry signal. An increase of the fundamental frequency from F_0 = 425 Hz to 440 Hz reduced the aversion threshold from 90 to 70 dB. As just verified by Zeskind and Marshall (1988) the variation of the fundamental frequency may convey different messages to the care-giving environment.

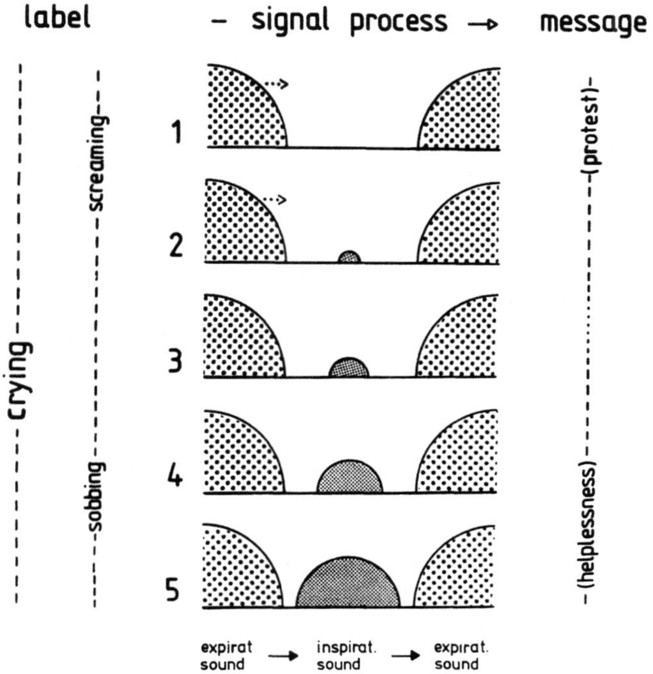

Fig. 9. Illustration of relationships between the conspicuousness of inspiratory sounds and interpretations of recipients given when judging a heard crying process. Besides the conspicuousness of inspiratory sounds, also their position has an impact on judgements. Changes in position may result from a shortening of the silent intervals preceding or following the inspiratory sounds

6 Conclusions and Perspectives

The responses which caregivers address to a crying infant depend on three variables: (1) prior information about the concerned infant, (2) information about the actual context of crying, and especially (3) cues from a particular crying process. The results given above have illustrated features of process dynamics and also their impact on caregiving behaviours. In the following, I will briefly focus on some aspects of our findings, and point out to perspectives for further investigations.

Internal state variables of signallers may be detected through an analysis of the acoustic features of their vocalizations (Hassenstein 1966, 1973; Todt 1968, 1977; Morton 1977; Scherer 1985, 1988). Crying processes may be a particularly good substratum for such endeavours: crying often covers a long time span, and, usually, is not interrupted by other behaviours. Thus, tonic components, such as affect and arousal and, especially, changes in these components may be encoded in the process dynamics. Nevertheless, it is not yet clear, whether and how gradual parameter changes, or parameter trends underlying such changes, are decoded by conspecific recipients.

A prominent feature of crying processes is the generally high variability of signal parameters. It has often been stated that a variable signalling may serve to prevent habituation in addresses. However, for the variability of crying processes, we have to consider an additional function which may arise from the irregular occurrence of extreme increases of signal parameters, such as the duration, the peaking amplitude or the fundamental frequency of a call (EPI's; SPP's). Basing on our results, I hypothesize that these events may act as a temporally distinct *trigger* of *caregiving responses*. Such a trigger could facilitate a caregiver's decision to look after an infant better than a just prolonged but monotoneous kind of signalling, and so could economically supplement the message of a crying encoded otherwise. As for the non-human primates, SPPs were found to temporally coincide with responses that often, such as a re-establishment of the body contact between an infant and his caregiver, terminated a crying (Fig. 6).

In human beings there is further evidence contributing to this issue. The evidence comes from aftereffects revealed in probands who, in earlier experiments, had been exposed to crying processes containing EPIs and SPPs with shrill acoustic features. In addition to other investigators, who also found that shrill crying features are an effective means of eliciting a caregiver's distress (Ostwald and Murry 1985; Bisping 1986), we had evidenced a conditioned distress response (Table 2). This aftereffect points to an interesting functional consequence of conspicuous changes of crying parameters: Since, after establishment of a conditioned distress response, EPIs and SPPs may elicit recipient distress even without containing shrill features. Thus the irregular performance of EPIs or SPPs may be interpreted as an economical interactional strategy. It may permit an infant to signal towards his caregiver successfully without crying extremely intensive (e.g. shrill) for a long time.

Recently, Gouzoules et al. (1984) have documented interactions among rhesus monkeys *(Macaca mulatta)* that were mediated by particular types of cries ("screams", in their terminology). In this species, juveniles apply specific screams as representational signals that, for instance, help a mother to recognize against which individual opponent the vocalizations of her offspring are addressed. The studied scream bouts of rhesus monkeys showed a clear homo-type structure. In contrast, the crying recorded from human infants and infants of Barbary macaques *(Macaca sylvanus)* often includes different cry types (Figs. 2, 3). This raises two questions that cannot be answered for Barbary macaque infants as yet: (1) Do the different cry types, occurring as consecutive process constituents, signal different messages? (2) Do the hetero-type crying processes of an infant provide an undifferentiated basis from which – during ontogeny – social learning would lead to genuine homo-type cry bouts, and also to the development of representational signals? Preliminary results have already shown that with the increasing age of infants, a remarkable decrease in the variability of acoustic and temporal crying features becomes apparent (Fig. 1).

Evidently, human infants, and likewise the young of non-human primates are able to determine how they can reach certain social goals by particular displays, and also how to shape their signalling behaviour accordingly. For crying, the learning aspect stresses two further questions: At which developmental stage may an infant become able to apply crying independently of its normal contextual variables (absence of mother; presence of a "harasser")? And, how far may an infant become able to cry without being in a genuine distress state (Todt et al. 1985; Whiten and Byrne 1986)?

Examinations of these questions may lead to a more advanced understanding of the communicative role of crying in primate infants. For comparative approaches, the "allomothering hypothesis" (p. 90) provides a promising perspective. And for studies focussing on the early expression and development of affectional-motivational variables, it seems promising, to supplement the analysis of static and dynamic crying features by considering also the reliable non-acoustic cues of this multi-directional display behaviour.

Acknowledgments. I am grateful to Ellen Merz and Baron Gilbert de Turckheim who gave permission to conduct our studies in "La forêt des Singes", Roc Amadour/France. In particular, I would like to thank Kurt Hammerschmidt and Bernhard Wolf for providing some of the data presented in this article, and Henrike Hultsch and Gottfried Wiedenmann for their comments on an earlier draft of the manuscript. Financial support from the Deutsche Forschungsgemeinschaft (grant: SPP Psychologische Ökologie; Multidirektionale Interaktionen; To 13/11) was greatly appreciated.

References

Ainsworth MDS (1973) The development of infant-mother attachment. In: Caldwell BM, Ricciuti HN (eds) Review of child development research, vol 3. Univ Chicago Press, Chicago

Ainsworth MDS, Bell SM, Stayton DJ (1974) Infant-mother attachment and social development: socialisation as a product of reciprocal responsiveness to signals. In: Richards MP (ed) The integration of a child into a social world. Cambridge Univ Press, Cambridge

Andrew RJ (1963) The origin and evolution of the calls and facial expressions of the primates. Behaviour 20:1–109

Baldwin JD, Baldwin JI (1976) Vocalizations of howler monkeys *(Alouatta palliata)* in southwestern Panama. Folia Primatol 26:81–108

Bell SM, Ainsworth MD (1972) Infant crying and maternal responsiveness. Child Dev 43:1171–1190

Bisping R (1986) Der Schrei des Neugeborenen: Struktur und Wirkung. Springer, Berlin Heidelberg New York Tokyo

Bosma JF, Truby HM, Lind J (1965) Cry motions of the newborn infant. Acta Paediatr Scand Suppl 163:61–92

Boukydis CFZ (1985) Perception of infant crying as an interpersonal event. In: Lester BM, Boukydis CFZ (eds) Infant crying. Plenum, New York London

Bowlby J (1969) Attachment and loss, vol 1. Attachment. Basic Books, New York

Brazelton TB (1962) Crying in infancy. Pediatrics 29:579–588

Brazelton TB (1985) Application of cry research to clinical perspectives. In: Lester BM, Boukydis CFZ (eds) Infant crying. Plenum, New York London

Brennan M, Kirkland J (1982) Classification of infant cries using descriptive scales. Infant Behav Dev 5:341–346

Brüser E (1985) Intradyadische Abstimmungsprozesse: Zur Entwicklung der frühen Face-to-face Interaktion. Diss, Freie Univ Berlin

Bryan YE, Newman JD (1988) Influence of infant cry structure on the heart rate of the listener. In: Newman JD (ed) The physiological control of mammalian vocalizations. Plenum, New York London

Chevalier-Skolnikoff S (1974) The ontogeny of communication in the stumptail macaque *(Macaca arctoides)*. Contrib Primatol 2:1–174

Deag JM, Crook JH (1971) Social behaviour and "agonistic buffering" in the wild Barbary macaque. Folia Primatol 15:183–200

DeVore I, Konner M (1974) Infancy in a hunter-gatherer life: an ethological perspective. In: White N (ed) Ethology and psychiatry. Univ Toronto Press, Toronto

Donnovan WL, Leavitt LA (1985) Physiology and behavior: parent's response to the infant cry. In: Lester BM, Boukydis CFZ (eds) Infant crying. Planum, New York London

Eisenberg JR (1976) Communication mechanisms and social integration in the black spider monkey, *Ateles fusciceps robustus*, and related species. Smithsonian Contrib Zool 213:1–108

Emde RN, Gaensbauer TJ, Harmon RJ (1976) Emotional expression in infancy: a biobehavioural study. Psych Issues 10 (I Monogr 37)

Epple G (1968) Comparative studies on vocalization in marmoset monkeys (Hapalidae). Folia Primatol 8:1–40

Fischer J (1988) Zur Frage der wechselseitigen Beeinflussung des Schreiverhaltens von Neugeborenen. Dipl Thes, Freie Univ Berlin

Fossey D (1979) Development of the mountain gorilla *(Gorilla gorilla beringei)*: the first thirty-six months. In: Hamburg DA, McCrown ER (eds) The great apes. Benjamin/Cummings, Menlo Park

Frodi AM (1981) Contribution of infant characteristics to child abuse. Am J Ment Def 85:341–349

Frodi AM (1985) When empathy fails: aversive infant crying and child abuse. In: Lester BM, Boukydis CFZ (eds) Infant crying. Plenum, New York London

Frodi AM, Lamb ME, Leavitt LS, Donovan WL (1978) Fathers' and mothers' responses to infant cries and smiles. Infant Behav Dev 1:187–198

Gautier J-P, Gautier-Hion A (1977) Communication in Old World monkeys. In: Sebeok TA (ed) How animals communicate. Indiana Univ Press, Bloomington

Golub HL, Corwin MJ (1982) Infant cry: a clue to diagnosis. Pediatrics 69:197–201

Gouzoules S, Gouzoules H, Marler P (1984) Rhesus monkey *(Macaca mulatta)* screams: representational signalling in the recruitment of agonistic aid. Anim Behav 32:182–193

Green S (1975) Variation of vocal pattern with social situation in the Japanese monkey *(Macaca fuscata)*: a field study. In: Rosenblum LA (ed) Primate behaviour: developments in field and laboratory research, vol 4. Academic Press, London New York, pp 1–102

Green S (1981) Sex differences and age gradations in vocalizations of Japanese and lion-tailed monkeys. Am Zool 21:165–184

Grimm RJ (1967) Catalogue of sounds of the pigtailed macaque *(Macaca nemistrina)*. J Zool 152: 361–373

Grossmann KE (1985) The development of emotional expression in a social context. In: Spence JT (ed) Motivation, emotion and personality. Elsevier, Amsterdam

Hammerschmidt K, Todt D, Kirkland J (1987) Crying behaviour in primate infants: cry call sequences. Cry Newslett 3:15

Harlow HF, Harlow MK (1965) The affectional systems. In: Schrier AM, Harlow HF, Stollnitz F (eds) Behaviour of nonhuman primates. New York London, pp 287–321

Hassenstein B (1955) Abbildende Begriffe. Zool Anz 18:197–202

Hassenstein B (1966) Kybernetik und biologische Forschung. Handbuch der Biologie, vol I/2. Akad Verlagsges Athenaion, Frankfurt/M

Hassenstein B (1973) Verhaltensbiologie des Kindes. Piper, München

Haut A, Todt D (1984) Early development of multidirectional interactions in *Macaca silenus*. Verh Dtsch Zool Ges 77:222

Hodun A, Snowdon CT, Soini P (1981) Subspecific variation in the long calls of the tamarin, *Saguinus fusciollis*. Z Tierpsychol 57:97–110

Hoffman M (1975) Developmental synthesis of affect and cognition and its implications for altruistic motivation. Dev Psychol 11:607–622

Hrdy SB (1976) The care and exploitation of non-human primate infants by conspecifics other than the mother. In: Rosenblatt J, Hinde R, Shaw E, Beer C (eds) Advances in the study of behaviour, vol 6. Academic Press, London New York

Hultsch H (1980) Beziehungen zwischen Struktur, zeitlicher Variabilität und sozialem Einsatz des Gesangs der Nachtigall *Luscinia megarhynchos B*. Diss, Freie Univ Berlin

Hultsch H (1983) Behavioural significance of duet interactions: cues from antiphonal duetting between males *(Cossypha heuglini H.)*. Behaviour 86:89–99

Hultsch H, Todt D (1982) Temporal performance roles during vocal interactions in nightingales *(Luscinia megarhynchos B.)*. Behav Ecol Sociobiol, pp 253–260

Hultsch H, Todt D (1986) Signal matching: Zeichensetzung durch mustergleiches Antworten. Z Semiot 8:233–244

Jürgens U (1979) Vocalizations as an emotional indicator. A neuroethological study in the squirrel monkey. Behaviour 69:88–117

Kummer H (1967) Tripartite relations in hamadrjas baboons. In: Altmann SA (ed) Social communication among primates. Chicago Univ Press, Chicago

Lawick-Goodall J van (1969) Mother-offspring relationships in free-ranging chimpanzees. In: Morris D (ed) Primate ethology. Doubleday, Garden City

Lee PC (1984) Early infant development and maternal care in free-ranging vervet monkeys. Primates 25 (1):36–47

Lester BM (1984) A biosocial model of infant crying. In: Lipsitt LP (ed) Advances in infancy research. Ablex, Norwood, pp 168–212

Lester BM (1985) There is more to crying than meets the ear. In: Lester BM, Boukydis CFZ (eds) Infant crying. Plenum, New York London

Lester BM, Boukydis CFZ (eds) (1985) Infant crying. Plenum, New York London

Lester BM, Zeskind PS (1979) The organization and assessment of crying in the infant at risk. In: Field TM, Sostek AM, Goldberg S, Shuman HH (eds) Infants born at risk. Spectrum, New York

Lester BM, Zeskind PS (1982) A biobehavioural perspective on crying in early infancy. In: Fitzgerald HE, Lester BM, Yogman MW (eds) Theory and research in behavioural pediatrics. Plenum, New York London, pp 113–178

Liebermann P (1985) The physiology of cry and speech in relation to linguistic behaviour. In: Lester BM, Boukydis CFZ (eds) Infant crying. Plenum, New York London

Liebermann P, Harris KS, Wolff P, Russell LH (1971) Newborn infant cry and nonhuman primate vocalization. J Speech Hear Res 14:718–727

Lieblich AK, Symmes D, Newman JD, Shapiro M (1980) Development of the isolation peep in laboratory bred squirrel monkeys. Anim Behav 28:1–9

Lillehei RA, Snowdon CT (1978) Individual and situational differences in the vocalizations of young stumptail macaques *(Macaca arctoides)*. Behav 65:270–281

Lind J (ed) (1965) Newborn infant cry. Acta Paediatr Scand 163

Lounsbury ML, Bates JE (1982) The cries of infants of differing levels of perceived temperamental difficultness: acoustic properties and effects on the listener. Child Dev 53:677–686

Marten K, Marler P (1977) Sound transmission and its significance for animal vocalizations: temperate habitats. Behav Ecol Sociobiol 2:271–290

Michelsson K, Sirvio P, Wasz-Höckert O (1977) Pain cry in full-term asphyxiated newborn infants correlated with late findings. Acta Paediatr Scand 66:611–616

Morton ES (1977) On the occurrence and significance of motivation-structural rules in some bird and mammal sounds. Am Nat 111:855–869

Murray AD (1979) Infant crying as an elicitor of parental behaviour: an examination of two models. Psychol Bull 86:191–215

Murray AD (1985) Aversiveness in the mind of the beholder: perception of infants crying by adults. In: Lester BM, Boukydis CFZ (eds) Infant crying. Plenum, New York London

Murry T, Hollien H, Mueller E (1975) Perceptual responses to infant crying: maternal recognition and sex judgements. J Child Lang 2:199–204

Newman JD (1985) The infant cry of primates: an evolutionary perspective. In: Lester BM, Boukydis CFZ (eds) Infant crying. Plenum, New York London

Newman JD, MacLean PD (1982) Effects on tegmental lesions on the isolation call of squirrel monkeys. Brain Res 232:317–329

Newman JD, Symmes D (1982) Inheritance and experience in the acquisition of primate acoustic behaviour. In: Snowdon CT, Brown CH, Petersen MR (eds) Primate communication. Cambridge Univ Press, Cambridge, pp 259–278

Nicolson NA (1987) Infants, mothers and other females. In: Smuts BB, Cheney DL, Seyfarth RM, Wrangham RW, Struhsaker TT (eds) Primate societies. Chicago Univ Press, Chicago

Oppenheimer JR (1977) Communication in New World monkeys. In: Sebeok TA (ed) How animals communicate. Indiana Univ Press, Bloomington

Ostwald PF (1963) Soundmaking: the acoustic communication of emotion. Thomas, Springfield

Ostwald PF (1972) The sounds of infancy. Dev Med Child Neurol 14:350–361

Ostwald PF, Murry T (1985) The communicative and diagnostic significance of infant sounds. In: Lester BM, Boukydis CFZ (eds) Infant crying. Plenum, New York London

Pola YV, Snowdon CT (1975) The vocalizations of pygmy marmosets *(Cebuella pygmaea)*. Anim Behav 23:826–842

Pook AGA (1977) A comparative study of the use of contact calls in *Saguinus fuscicollis* and *Callithrix jacchus*. In: Kleiman DG (ed) The biology and conservation of the Callitrichidae. Smithsonian Inst, Washington DC

Rowell TE, Hinde RA (1962) Vocal communication by the rhesus monkeys *(Macaca mulatta)*. Proc Zool Soc London 138:279–294

Sagi A (1981) Mothers' and non-mothers' identification of infant cries. Infant Behav Dev 4:37–40

Seyfarth RM, Cheney DL (1982) How monkeys see the world. In: Snowdon CT, Brown CH, Petersen MR (eds) Primate communication. Cambridge Univ Press, Cambridge

Scherer KR (1985) Vocal affect signalling. In: Rosenblatt JS, Beer C, Busnel MC, Slater PJ (eds) Adv Study Behav 15:188–244

Schlittgen R, Streifberg BHJ (1987) Zeitreihenanalyse. Oldenbourg, München

Simons RC, Bielert CF (1973) An experimental study of vocal communication between mother and infant monkeys *(Macaca nemestrina)*. Am J Phys Anthropol 28:455–462

Snowdon CT, Hodun A (1981) Acoustic adaptions in pygmy marmoset contact calls: location cues vary with distance between conspecifics. Behav Ecol Sociobiol 9:295–300

Snowdon CT, Cleveland J, Frensch JA (1983) Responses to context- and individual-specific cues in cotton-top tamarin long calls. Anim Behav 31:92–101

Struhsaker TT (1967) Auditory communication among vervet monkeys *(Cercopithecus aethiops)*. In: Altman SA (ed) Social communication among primates. Chicago Univ Press, Chicago

Suomi SJ (1977) Adult male-infant interactions among monkeys living in nuclear families. Child Dev 48:1255–1270

Taub DM (1980) Testing the "agonistic buffering" hypothesis: the dynamics of participation in the triadic interaction. Behav Ecol Sociobiol 6:187–197

Tembrock G (1971) Biokommunikation. Akademie, Berlin

Thorpe WH (1972) Duetting and antiphonal songs in birds. Its extent and significance. Behav Suppl XVIII:1–197

Todt D (1968) Zur Steuerung unregelmäßiger Verhaltensabläufe. In: Mittelstaedt H (ed) Kybernetik. Oldenbourg, München, pp 465–485

Todt D (1973) Biologisch-kybernetische Analyse der Komposition des Gesangs verschiedener Vögel. Nova Acta Leopold 208:311–331

Todt D (1975) Short term inhibition of outputs occurring in the vocal behaviour of blackbirds *(Turdus merula m.L.)*. J Comp Physiol 98:289–306

Todt D (1977) Zur infradianen Rhythmik im Verhalten von Vertebraten – Ergebnisse aus Analysen des variablen Verhaltens von Singvögeln. Nova Acta Leopold 225:607–619

Todt D (1986) Hinweischaracter und Hinweisfunktion von Verhalten. Z Semiot 8:183–232

Todt D (1988) Duels, duets, dialogues. Behaviour (in press)

Todt D, Hultsch H (1980) Functional aspects of sequence and hierarchy in bird song. Acta XVII Congr Int Ornithol, Berlin, pp 663–670

Todt D, Pohl R (1984) Communicative strategies in estrous Barbary ape female *(Macaca sylvanus L.)* during copulation behaviour: advertising, triggering, offlimiting. Verh Dtsch Zool Ges 77: 225

Todt D, Hultsch H, Duvall II FP (1981) Behavioural significance and social functions of vocal and non-vocal displays in the monogamous duet-singer *Cossypha heuglini*. Zool Beitr 27:421–448

Todt D, Brüser E, Hultsch H, Lange R (1982) Nocturnal actions and interactions of newborn monkeys. J Hum Evol 11:383–389

Todt D, Hammerschmidt K, Pohl R (1985) The dusk-chorus of the Barbary macaque *(Macaca sylvanus)*: an age-specific performance. Verh Dtsch Zool Ges 78:217

Truby HM, Lind J (1965) Cry sounds of the newborn infant. In: Lind J (ed) Newborn infant cry. Acta Paediatr Scand Suppl:163

Wasz-Höckert O, Lind J, Vuorenhoiski V, Partanen T, Valanne E (1968) The infant cry. In: Clinics in developmental medicine. Lavenham, London

Whiten A, Byrne R (1986) The St Andrews catalogue of tactical deception in primates. St Andrews Psycholog Reports 10:1–47

Whitten PL (1987) Infant and adult males. In: Smuts BB, Cheney DL, Seyfarth RM, Wrangham RW, Struhsaker TT (eds) Primate societies. Chicago Univ Press, Chicago

Winter P, Handley P, Ploog D, Schott D (1974) Ontogeny of squirrel monkey calls under normal conditions and under acoustic isolation. Behaviour 47:230–239

Wolff A (1985) Zum bioakustischen Verhaltensrepertoire neugeborener Berberaffen *(Macaca sylvanus)* mit Bezug auf den situativen Kontext. Dipl Thes, Freie Univ Berlin

Wolff A, Todt D (1985) The effect of vocal signals performed by newborn monkeys *(Macaca sylvanus)* on infant-caretaker-distances. Verh Dtsch Zool Ges 78:219

Wolff P (1969) The natural history of crying and other vocalizations in infants. In: Foss BM (ed) Determinants of infant behaviour IV. Methuen, London

Wolffgramm J, Thimm F (1976) Bearbeitergesteuerte Analyse von Verhaltenszeitreihen am Digitalrechner. Biol Cybern 21:61–88

Wolffgramm H, Todt D (1982) Pattern and time specificity in vocal responses of blackbirds *Turdus merula L.* Behaviour 81(2–4):264–286

Zeskind PS (1980) Adult responses to cries of low and high risk infants. Infant Behav Dev 3:167–177

Zeskind PS, Marshall TR (1988) The relation between variations in pitch and maternal perceptions of infant crying. Child Dev 59:193–196

Communications as Social Interaction: Its Importance in Ontogeny and Adult Behavior

CHARLES T. SNOWDON[1]

Much of our current understanding of primate communication is based on the analysis of individual signals, on the sequences of signals given by an individual, and on the responses of animals to playbacks of these signals. In reality, however, communication is a social interaction with individuals constantly altering their signals in response to the history of the immediate interaction and in response to the history of the social relationships between individuals. The ontogeny of vocal communication is also dependent upon the social interactions between young animals and other group members. The studies discussed here describe turn taking in adult pygmy marmosets and on the role of past history of social relationships in their perceptual responses to calls. Data are presented on the complex vocal interactions that occur within and between troops of cotton-top tamarins during territorial encounters. In both turn taking and in territorial encounters there appear to be some general rules organizing calling patterns. Finally, the role of social imitation and social environment on the ontogeny of vocalizations in marmosets and tamarins will be discussed.

1 Complex Models for Communication

The early model for studies of communication was quite simple, focusing on the communicator, the signal produced by the communicator, and the response made by the recipient. In reality, though, communication is a much more complex system. There are at least three types of complications that should receive attention.

First, communication does not take place as an isolated instance of a signal passing between communicator and recipient. The previous history of the relationship between the communicator and the recipient will affect the signal selected to be used in a given situation and will affect the response to the signal given by the recipient. The role of prior social history is readily acknowledged in studies of the social organization of primate groups where the past history of dominance-subordinance relations, kinship systems, and other social alliances are all known to be important. However, there are few studies on the role of these relationships on the communicative behavior of primates (see Seyfarth and Cheney 1986 for an exception).

Second, a phenomenon basic to human communication is the conversation. We rarely produce utterances in isolation or, with the exception of the classroom and the scientific conference, as monologues. Instead, the socially natural unit of communica-

1 Dept. of Psychology, University of Wisconsin, 1202 West Johnson Street, Madison, WI 53706, USA

Todt/Goedeking/Symmes (Eds.)
Primate Vocal Communication

tion is not the phoneme, the word, or the sentence, but the conversation. Communication is truly an interactive process, yet both in human linguistics and in animal communication we have confined ourselves most frequently to the analyses of single utterances. There is good reason for this. We can imagine many more decades of research being necessary before we fully understand even a few simple utterances. Yet, I would like to suggest that we should start developing some techniques for analyzing vocal interactions (see chapters by Maurus et al. and by Simmes and Biben, this Vol.). Ultimately, the comprehensiveness of our understanding of animal communication or human language will require the understanding of the social interactive processes of communication, not simply the understanding of simple, single utterances.

Third, communication is frequently not restricted to the interactions between two individuals, but is a multipartite process. An animal may produce one signal for one individual but different signals for other individuals (see Marler and Mitani, this Vol., for a discussion of audience effects). In addition, different individuals hearing the same utterance may respond to the utterance in different ways, and how they respond will have differing effects on the communicator which will affect the nature of the communicator's next utterance.

Each of these three complications to our traditional model of communication will make our task of understanding vocal communication considerably more difficult, but only when communication is understood in its proper social context will we have a real understanding of communication.

In the rest of this chapter I will present the results of our initial attempts to study the importance of vocal interaction. I will first review studies on adults, describing conversational patterns found among pygmy marmosets and how the history of social interactions affects the perception of their calls. Then I will discuss the complex vocal interactions found among cotton-top tamarins during intertroop vocal interactions. The last part of the chapter focuses on our recent studies on ontogeny. In pygmy marmosets we find that there are structural changes that occur progressively over development with final adult structure being achieved before proper adult usage of calls. I will report on vocal regression where older siblings regress to infantile vocalizations on the birth of new infants in the group. In cotton-top tamarins I will describe some early analyses of structural changes in chirp vocalizations comparing the results of hand-reared and normally-reared infants. Finally, in a parallel with bird song, I will present data on changes on vocal signaling that result when animals reach sexual maturity, a state that is controlled in tamarins and marmosets by social environment, not by chronological age or physical maturation. Although few of these results are definitive as yet, they do suggest some ways in which we can start to develop some new directions in the study of primate vocal communication.

2 Interactional Communication Among Adults

We have studied two forms of interactional communication in the pygmy marmoset *(Cebuella pygmaea)* both of which involve the contact trills used by animals as they move through their environment. These calls were first described in captive animals by Pola and Snowdon (1975) and subsequently observed in the wild (Snowdon and Hodun

1981). The three forms of trills (quiet trills, closed-mouth trills, and J-calls) were used by animals to keep in contact with one another as they moved through the home range. They were used to recruit animals to sleeping sites at night and were used by sentinel animals while others groomed and sunned themselves. The calls are individually distinct, and playback studies indicated that animals respond individually to specific calls (Snowdon and Cleveland 1980).

The first study involved the use of calls spontaneously in groups (Snowdon and Cleveland 1984). If animals use calls to keep in touch with each other, then, we thought, it is likely that there might be some regular pattern of calling so that all animals in a group can keep track of one another. We observed a captive group of three animals and recorded sequences of these contact calls, a total of 295 bouts and 1315 sequences of three calls each. Using information theory analysis we discovered that there was considerable predictability in call sequences. Knowledge of which animal had called previously and which two animals had called previously both provided predictability about which animal would call next. There were orderly patterns to the sequences of calls. Most notably, there was a significant tendency to take turns. That is, each of the three animals would call in sequence much more often than would be expected by chance, and an animal would interrupt a sequence by calling twice in succession considerably less often than would be predicted by chance. One particular sequence of calling of three animals (1, 2, 3) occurred significantly more often than the other sequence of

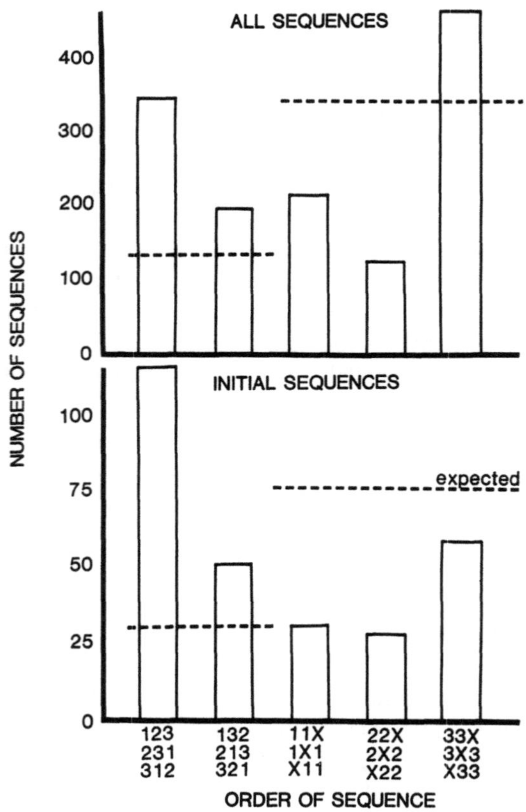

Fig. 1. Sequences of calls given in a group of three pygmy marmosets. *Top panel:* all sequences of calling. *Lower panel:* initial sequences of calls in a bout. The number of sequences for each animal calling once within three calls is greater than expected by chance (Snowdon and Cleveland 1984)

three animals (1, 3, 2) (Fig. 1). In human children the development of turn taking is considered to be a skill of high cognitive achievement (passing from egocentricity to the ability to understand another individual's perspective). It is all the more remarkable that a small primate like the pygmy marmoset is able to exhibit turn-taking behavior.

The second example of interactional influences on communication in pygmy marmosets relates to how monkeys perceive their calls. Several years ago we showed that pygmy marmosets responded in a categorial fashion to two trills that differed from each other only in duration (Snowdon and Pola 1978). The closed-mouth trill was used in situations of calm movement throughout the colony. The open-mouth trill differed from the closed-mouth trill only in duration. However, the open-mouth trill was accompanied by a significant increase in aggressive and fearful behavior in contrast to the closed-mouth trill. We developed a pygmy marmoset trill synthesizer and in a playback study showed that the monkeys had a sharp boundary in their response to these two calls. All durations of trills up to 250 ms were responded to as though to a natural closed-mouth trill, while all calls longer than 250 ms were responded to at the same levels as background noise. A 9-ms difference between stimuli on either side of the 250-ms boundary produced different responses.

While the functional significance of a call boundary was clear, the significance of failing to discriminate between different closed-mouth trills was considerably less clear. It is odd that animals should neglect information in the call and treat quite different calls with equivalent responses. Thus, pygmy marmosets responded equivalently to synthesized trills of 176 and 248 ms but quite differently to calls of 248 and 257 ms duration. Why should this be? One possibility is that animals do perceive differences in duration within a call category, but that the testing situation was not appropriate to determine whether the animals perceive the differences. Thus, if we asked a human subject "Do you hear a /ba/?" we might be told that two very different /ba/'s were the same. If we were to ask if the same individual produced both /ba/'s, then we might get a response that the two calls were different. We attempted to do a similar experiment with pygmy marmosets.

We recorded the closed-mouth trills of five new pygmy marmosets. Two pairs of animals among the five had very similar closed-mouth trills to one another, so we took the average stimulus for each of those pairs. After we collected a large sample of trills from each of the two pairs and the single individual, we measured the parameters of the calls and found individual differences between the calls on most of the acoustic variables we measured. We then programmed our synthesizer to produce calls that imitated the individual specific features of each of the three groups and then systematically varied the duration of these calls in 100-ms increments from 100 to 600 ms. We observed (Snowdon and Cleveland 1980) that when we played back the call of an individual through a speaker located in that animal's cage, then other animals in the colony would give antiphonal calls. However, when we played back the same stimulus through a speaker in a cage in another part of the colony, the animals failed to respond antiphonally. In the present study we played synthesized calls an equal number of times through a speaker in the stimulus animal's home cage and from a speaker in another part of the environment. A significant difference in the response to a call played back in the stimulus animal's home cage and the same call played back from a different part of the colony was used to infer individual recognition. If we failed to find a dif-

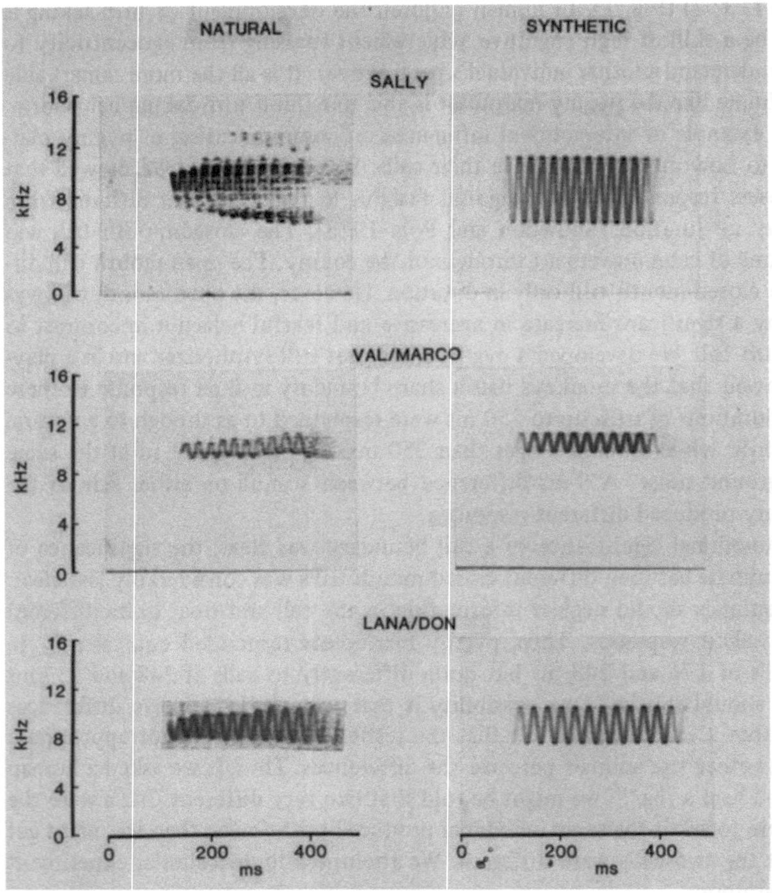

Fig. 2. Examples of natural and synthesized closed-mouth trills from three individual/pairs (Snowdon 1987)

ferential response then that particular stimulus would not have been perceived by the other animals as belonging to a familiar individual (seee Snowdon 1987, for more details). Figure 2 presents examples of natural and synthesized trills from each of the three stimulus groups (Sally, Val/Marco, and Lana/Don). Figure 3 presents the responses to all stimulus trials. The pattern of significant responses closely parallels the distribution of differences in call structure shown in the histograms in the lower part of the figure. Figure 4 presents responses to the calls of one animal, Sally. Sally had a very narrow distribution of call durations, and the responses of other animals to Sally's calls showed an equally narrow distribution. Instead of a single categorical response function, there are different response functions correlated with the pattern of duration distributions of individual animals.

In our previous study where a 9-ms difference in call duration produced different responses we had hypothesized the existence of neurological feature detectors. From the second study, it is clear that a hard-wired system cannot account for the results.

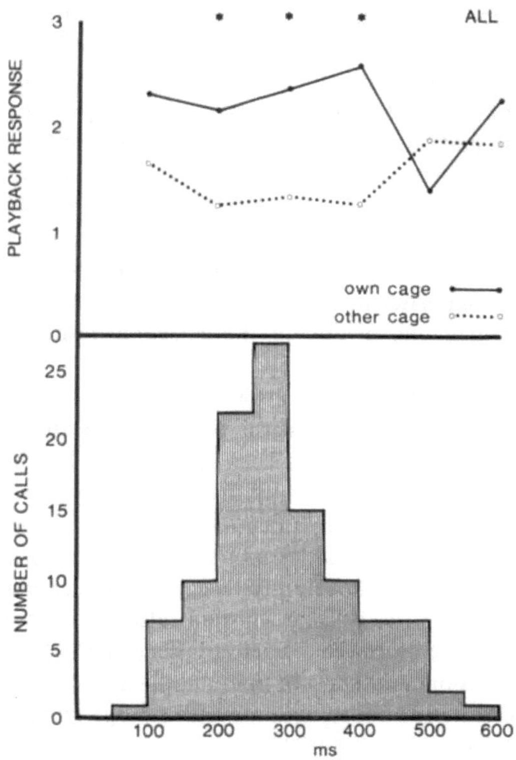

Fig. 3. *Top:* responses to playback of synthesized calls of all stimulus animals from cage location of animal whose call is being played versus from another location (* $p < 0.05$). *Bottom:* distribution by duration of spontaneous closed-mouth trills from all animals (Snowdon 1987)

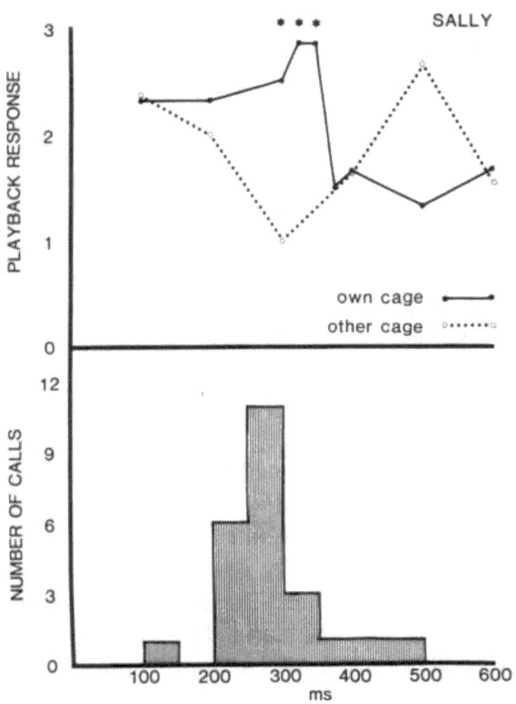

Fig. 4. *Top:* responses to playback of synthesized calls of Sally from Sally's own cage versus from other cages (* $p < 0.05$). *Bottom:* distribution by duration of Sally's closed-mouth trills (Snowdon 1987)

Obviously, animals make use of their previous social experience with familiar individuals and construct phonetic boundaries appropriate to this past social experience. There may be a feature detector that determines the upper boundaries for responses to calls from novel animals, but we must posit a flexible phonetic boundary system capable of being modified by social experience with particular individuals in order to account for the results of the second study. Thus, previous social interactions with familiar animals can affect the nature of perceptual responses. Much of the strong evidence for categorial perception that has been obtained in both human and animal studies may be due to the use of averaged synthetic sounds that represent no particular familiar individual. If instead subjects are presented with stimuli that match in all other dimensions the characteristics of socially familiar individuals, then the rigidity of the categorial response system might disappear (Snowdon 1987).

The final example of interactional communication in adults is from another species with which we have worked extensively, the cotton-top tamarin *(Saguinus o. oedipus)* an endangered species from northern Colombia. There have been few field studies on this species and its closely related subspecies, Geoffroy's tamarin *(Saguinus o. geoffroyi)* (Neyman 1977; Dawson 1979). The species is highly territorial living in small groups, and animals encounter other groups at territorial boundaries about once or twice each week. The typical response is for animals to engage in extensive vocal communication including large numbers of long calls, two- or three-syllable whistled calls that are between 2-3 s in duration. In captivity we have described several variations on the long calls: one variant used for intragroup cohesion, another two variants, the normal long call and the F+ whistle call, being used in territorial encounters with strange groups. The F-chirp and the F-chirp trill are also used in territorial encounters or intergroup conflict situations (Cleveland and Snowdon 1982).

Territorial encounters are difficult to study in the field, but the descriptions from the field indicated that such encounters might be very fruitful places in which to study vocal interactions between groups. In captive colonies, one can record vocalizations simultaneously from two different groups, while observing which individuals within a group are giving the vocalizations. McConnell and Snowdon (1986) devised a simple method for stimulating territorial vocal interactions. We reasoned that one characteristic of the start of a territorial encounter in the wild would be that two groups would move toward each other unaware until they began hearing increasingly louder intragroup vocalizations from the other group. To mimic this phenomenon in captivity we simply opened the doors between two different colony rooms which produced a 5-dB increase in the intensity of intragroup vocalizations. We placed microphones in each of the rooms so that we could keep track subsequently of the sequence of vocalizations within and between rooms. The monkeys did not respond immediately to the door opening, but within 30 s one or two animals in a group would look toward the direction of the other room as if they had just detected the strange group. They would begin to give an array of long calls and F-chirps. These calls were quickly answered by the other group and a complex vocal exchange continued for the duration of the 12-min observation period. Similar observations conducted when the doors were briefly opened and then shut served as control trials. The animals demonstrated behavioral changes predicted from observations in the wild. There was an increase in scent marking and piloerection and a decrease in huddling in the experimental conditions compared with the control conditions. There was an increase in all of the intertroop vocalizations described by Cleveland and Snowdon (1982). Thus, the simple manipulation of opening doors between different colony rooms had the expected effect of stimulating vocal exchanges similar to those induced by playbacks in the wild (see Marler and Mitani, this Vol.).

The more interesting results are from the analyses of the vocal interactions. There were sex and age differences in calling. The participants in the long calling bouts were primarily paired animals which had no offspring. The structure of all the territorial vocalizations given were characteristic of sexually mature animals (see below). There was evidence of role differentiation. Females gave significantly more long calls than would be expected by chance, while males gave significantly more F-chirps and F-chirp trills than expected. However, the F+ whistle call, which is a syntactic combination of the F-chirp and the long call, was given with equal frequency by both sexes.

Close analyses indicated that responses were not given randomly. In general, intergroup territorial calls tended to follow one another and affiliative calls tended to follow one another. There was not a rapid switching back and forth between affiliative and territorial calls. Different responses were given to a call depending on whether the response was from within the same group as the focal caller or from the other group.

A diagram of the response to focal callers is shown in Fig. 5. The left-hand side shows the responses given by animals in the same room as the focal caller, while the right-hand side shows the responses given by animals in the other room. There was an escalation of response from F-chirp to F-chirp trill or to F+ whistle calls given by animals within the same room as the focal caller. In contrast, animals in the other room tended to respond with the same type of call as the focal caller. There are different communicative roles played by different sexes and different responses given according to whether the last call one heard was from one's own group or from the other group. In general, there appears to be an inciting within the group to more and more intense versions of intertroop calling and an imitation or even a softening of response with some affiliative calls given between groups. The vocal interactions observed between groups are not random.

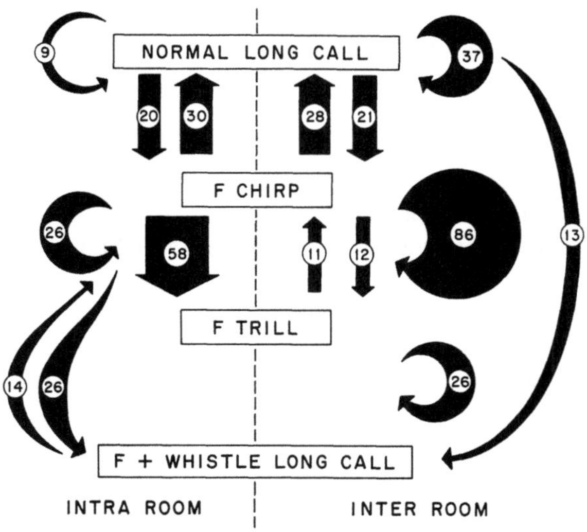

Fig. 5. Vocal interactions during long call bouts in cotton-top tamarins. *Left side* displays transitions from call types between callers within the same group. *Right side* displays transitions between call types between callers in different rooms (McConnell and Snowdon 1986)

The techniques were have used so far to analyze the patterns of vocal interaction involve rather simple manipulations and observations, but I think they hold considerable promise for future study of interactive communication.

3 Importance of Interaction in Ontogeny

Understanding the interactive process of communication is essential for dealing with the major underdeveloped research problem in primate communication – ontogeny. To date, our information on the ontogeny of vocal communication is quite limited. Our major data are from the squirrel monkey, primarily studies of infant isolation peeps, and the major finding is that call structures are inherited, not learned (Newman and Symmes 1982). This is a paradox, since we have some clear parallels in the development of bird song and the development of human speech (Marler 1970b).

Why should we not find any parallels in the work on primates? This might be due to several reasons. First, research has tended to focus on a single call type, the isolation peep, which is used by infants separated from their mothers. This is one type of call that might be expected to be selected as a highly conservative call. Alarm calls and distress calls to be effective should not be subject to great degrees of developmental plasticity. Thus, it might be more appropriate to look at other call types for evidence of modifiability during development. Second, only a few species have been studied, and in birds many precocial species show little or no evidence of modifiability, while most altricial species do show a need for early experience in vocal development. Even among altricial species, however, there is a great variety with some species appearing to have a critical period for song learning while other species are able as adults to learn components of their mates' songs (Mundinger 1970) or to learn the sounds of another species (Pepperberg 1986). Third, we might not be applying the proper models to the study of ontogeny.

Chomsky (1957) proposed that no behaviorist theory could account for the complexity of human language and its capacity for the generation of an infinite number of sentences. Chomsky argued that there must be several innate mechanisms to assist in language acquisition. At the same time the research of ethologists had an increasing influence on thinking in North America, and the early research on bird song learning by Thorpe (1958) and Marler (1970a,b) suggested an ontogenetic model compatible with Chomsky's ideas. There was a species-specific predisposition to acquire only a certain limited range of song types. These songs, however, had to be presented to birds during a critical period in development. Thus, there was a need for learning, but the timing of learning and what was learned were constrained by species-specific mechanisms.

In parallel, language acquisition was said to be limited by the range of potential phonemes available to human speakers and by grammatical universals. An infant needed only minimal exposure to sounds since children born to deaf parents would learn language readily with no signs of deficit, and language learning was limited to a critical period since children isolated from language exposure for the first 12 years of life failed to learn language (Lenneberg 1969). Studies on newborn infants indicated a predisposition to categorize speech sounds in the same way as adult speakers did, and children could even make perceptual distinctions important in a different language that their parents could no longer make.

Recent work on both child language and bird song suggests a modification of this earlier, ethologically-based view. The importance of close social interaction is seen as a key to language acquisition. The various games and rituals between caregivers and human infants create a context for vocal communication and provide a structure for the acquisition of syntactic rules (Bruner 1983; Snow and Ferguson 1977). Babbling in infants plays an important role in phonetic development (e.g., Oller et al. 1975), and babbling can be stimulated through social and vocal reinforcement from adults (Dodd 1972; Todd and Palmer 1968). Parents use a simplified language in communicating with infants which may serve to provide a simplified model for infants to attain linguistic competence (Ferguson 1964; Ferguson and Garnica 1977).

Interactive models of vocal acquisition in bird song are becoming more common. The presence of a live tutor has been used to extend the period of song learning beyond the usual sensitive period, and parts of allospecific song can be acquired with a live tutor (Todt 1975; Todt et al. 1979; Baptista and Morton 1981; Baptista and Petrinovitch 1984; Kroodsma 1972, 1973; Kroodsma and Baylis 1982; Pepperberg 1986). Marler and Peters (1982) have demonstrated the importance of a period of vocal practice (or plastic song) that appears similar in function to the babbling of infants. Lemon (1975) discusses the importance of vocal improvisation in acquiring song dialects. Thus, it appears that vocal learning in human beings and in birds is not only more flexible than previously thought but also is highly dependent upon social interaction with other organisms and upon a period of vocal play or practice.

Despite the essentially negative results with squirrel monkey isolation peeps, there are some recent studies that indicate more openness in primate vocal ontogeny. Seyfarth and Cheney (1986) showed that young vervet monkeys do not use their alarm calls in as refined a way as adults. They typically give aerial predator calls to a variety of aerial species not only to those that prey on vervets. However, adults appeared to reinforce young vervet monkeys when the young animals gave an aerial predator call in the presence of true predators. Infant vervets did not show the differentiated structure of grunts seen in adults (Cheney and Seyfarth 1982). Initially the calls are more variable with different parameters reaching adult parameter values with different time courses. Grunt vocalizations were not differentiated with respect to context as completely as adult grunts until the infants were 3- to 4-years-old. The general picture of vervet vocal ontogeny is that calls are highly variable with only limited context specificity in the early months. With practice and social learning both from observation and from socially reinforced behavior, infants develop a better usage of calls, and they are able to produce calls with a structure close to that of adults.

We have studied the role of social interactions in vocal development in three different systems: the development of trills in pygmy marmosets, the differentiation of chirp types in cotton-top tamarins, and the ontogeny of long call usage in cotton-top tamarins.

4 Trill Development in Pygmy Marmosets

C. Sweet and I have studied the ontogeny of the trill vocalizations in pygmy marmosets. We focused on the most common of the trill variants, the closed-mouth trill. This call was found in young monkeys in our earliest samples at 3 weeks of age. However, there were several differences in the structure of these early closed-mouth trills when compared with the adult form. The earliest trills were much higher in pitch, were much shorter in duration, and the frequency envelope was much more irregular (that is, the upper and lower boundaries of the frequency envelope varied greatly over the course of the call) than with adult calls. The three parameters of pitch, duration, and regularity of frequency envelope approached adult levels with independent time courses. As seen in Fig. 6 pitch reached adult levels by the sixth week of life. The call duration increased to the range of adult durations somewhere between the 10th and 15th week of age, while the percentage of calls with irregularity of the frequency envelope showed a gradual decline toward adult levels over the entire 40 weeks of the study. Coincident with the observations of structural change, we also observed the context in which the calls were used. Typically, adults use closed-mouth trills as affiliative contact or location calls, and the trills are never found in conjunction with other types of vocalizations. In contrast, almost all of the infant closed-mouth trills were found in conjunction with other types of calls, and the trills appeared to be uttered indiscriminately without regard for the proper context. Gradually, the appropriate usage of the calls changed to adultlike usage, but not until week 38 of our sampling did any of the pygmy marmosets show both adult structure and adult usage of the trills.

An interesting finding shown in Fig. 6 is a vocal regression that occurred between weeks 28 and 30. At this time the duration of the trills became much shorter and the pitch became much higher, regressing to the levels found in the very youngest infants. This period coincided with the birth of new infants in each of the groups we observed,

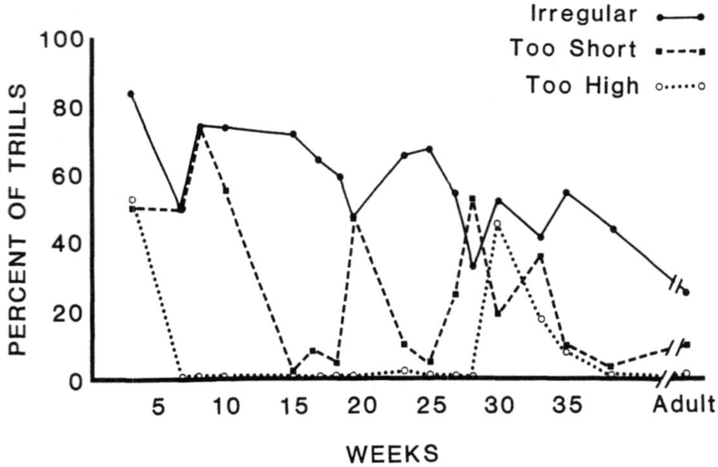

Fig. 6. Changes in structure of closed-mouth trills during first 40 weeks of life in pygmy marmosets. Note the regression to shorter and higher pitched calls that occurs at about 30 weeks of age when new infants were born (Snowdon 1988)

and the regression in the form of infantile vocalizations appeared to be the response of the 30-week-old animals to the loss of attention resulting from the birth of new infants.

The change in the pitch of the call can be related to physical changes in the body size that would result in lower pitch. However, the rapidity of the change in pitch within the first 6 weeks is much earlier than the major growth of pygmy marmosets and well before puberty. The alterations in trill duration and in the reduction of irregularity of the frequency envelope cannot be related easily to simple maturational changes. The independence of change in each of these parameters and the regression toward infantile call forms with the birth of new infants suggests some flexibility in vocal development which, while not a demonstration of call learning, implies at least some developmental flexibility in call development. The independent changes in call parameters and the failure to use the calls in appropriate context until relatively late in infancy closely parallel results reported for vervet monkeys by Seyfarth and Cheney (1986).

We have some preliminary data on the ontogeny of the J-call, another form of trill vocalization. This form appears later in infancy (between week 14 and week 22 in normally-reared infants) and the J-calls do not appear to be as highly variable as the early closed-mouth trills. However, J-calls at this age are given in inappropriate contexts just as the closed-mouth trills.

5 Chirp Development in the Cotton-Top Tamarin

In the adult cotton-top tamarin Cleveland and Snowdon (1982) demonstrated eight variants of chirps that could be discriminated statistically from one another and which appeared to be used in very different contexts. L. Dronzek and I have studied the ontogeny of chirp vocalizations in cotton-top tamarin infants throughout the first year of life. Twice weekly vocal samples were taken of several infants. It seemed likely that the appearance of some chirp variants would occur in the first months of life. In our vocal recording we took careful note of the context in which the calls were given. We measured the same vocal parameters of infant chirps that had been used to differentiate among the adult chirps and then compared these parameters in chirps given in different contexts by infant cotton-top tamarins. Surprisingly, we could find no consistent differences between chirps uttered in different situations, even in 12-month-old infants. Thus, the differentiation of chirp structure by context did not appear in the first year of life in cotton-top tamarins.

The chirps, although undifferentiated by context, were quite variable when compared with adult structure. The coefficients of variation of three parameters of chirps-start frequency, end frequency, and duration over the first year of life were much higher than those for adult calls. As with the pygmy marmoset closed-mouth trills there were different time courses for different variables. End frequency reached adult levels of variation by the fourth to sixth month. Start frequency showed a gradual reduction in the coefficient of variation over the course of the first year but had not reached adult levels by the end of the year. Duration was within the adult range by the fifth month but displayed considerable variation.

6 Long Call Development in Cotton-Top Tamarins

We have developed techniques for measuring estrogens, testosterone, and luteinizing hormone/chorionic gonadotropin in the urine of cotton-top tamarins. We have measured hormonal levels on a daily basis in male and female cotton-top tamarins while living as subordinate members within a family group and after being removed from the group and paired with a mate in another colony room. The hormonal changes are quite striking. While we have found occasional examples of estrogen or LH activity in females living in the family, we have seen no evidence of orderly cycles indicative of ovulation. In contrast, within as few as 8 days after being removed from the family, mated females have been observed to ovulate, and they often conceive on the first or second ovulation (Ziegler et al. 1987). The social environment in which tamarins live has a profound effect on reproductive physiology.

These different social environments also play an important role in vocal development. The major vocalizations used in territorial disputes are the normal long call, the F-chirp, the F-chirp trill, and the F+ whistle call (see Cleveland and Snowdon 1982 for a description of each of these calls). An additional vocalization, the quiet long call, is used for intragroup cohesion, and is given at a high rate when a mated pair is temporarily separated from one another. Figure 7 shows that each of these calls appeared with extremely low frequency in animals living in a subordinate role within the family group, but each of the calls occurred at significantly higher rates after the animals were removed from the family and mated. Thus, there is a close correlation between the development of adult reproductive competence and the appearance of these vocalizations. We do not yet know whether there is a causal relationship between hormonal changes and the appearance of the vocalizations or whether both change independently as a function of the change in an animal's social environment. However, there are close parallels between the appearance of these calls with reproductive maturity and the appearance of bird song with male reproductive maturation (Marler and Peters 1982). There is also a highly variable form of long call, the combination long call, which has many of the characteristics of subsong and plastic song described by Marler and Peters (1982) as occurring prior to the attainment of reproductive maturity in birds. These earlier phases of song development in birds are thought to represent a period of practice and crystallization of song learning in birds. The close parallels between bird song onto-

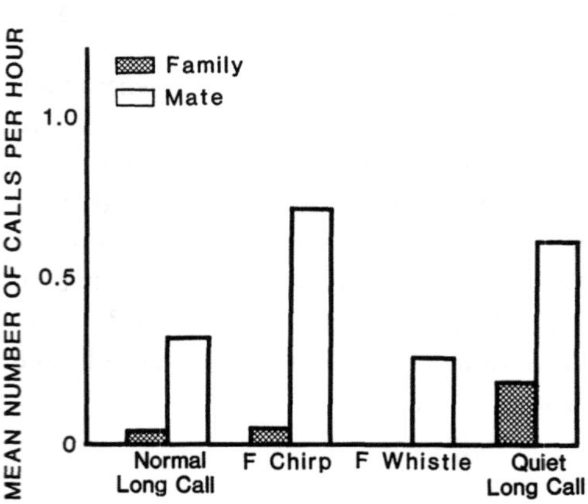

Fig. 7. Number of calls per hour in cotton-top tamarins while living in family groups and in the first 4 months after being removed from the family and paired with a mate. Normal long calls, F-chirps, and F-whistles are used in territorial encounters. Quiet long calls are used as cohesive calls between mates (Snowdon 1988)

geny and the development of adult forms of long calls in cotton-top tamarins suggests the possibility of a similar ontogenetic process for long call ontogeny.

7 Conclusions

While much progress has been made in the study of primate communication over the last 2 decades, two major related areas remain relatively unexplored: the use of vocalizations in complex social interchanges between animals and the ontogeny of vocalizations. Monkeys have conventions for turn taking in their calling; there are orderly and predictable patterns in the types of calls exchanged within and between troops of monkeys during territorial encounters and an animal's previous social experience with given individuals can influence how that animal perceives and responds to synthesized calls mimicking that individual. These studies provide the beginning of a framework for studying the roles of social history and interaction on communication. The social interactions between caregivers and infants play an important role in the ontogeny of human speech and language. Similar social influences should be examined in the ontogeny of primate communication. There are orderly changes in the development of both the closed-mouth trills of pygmy marmosets and the chirps of cotton-top tamarins with different vocal parameters having different developmental time courses. These different time courses, the presence of vocal regression with the birth of new infants, the lack of correct context of usage when adult structure has been achieved, and similar findings in vervet monkeys (Seyfarth and Cheney 1986) suggest that vocal development in these species may be more plastic and flexible than in many of the monkey species that have been studied to date. The parallels in long call production and other territorial calls with hormonal maturation in tamarins compared with song development and hormonal maturation in birds suggest that further study of long calls in tamarins might be a fruitful source of information on the vocal ontogeny in primates.

Acknowledgments. The research of my laboratory was supported by USPHS grants MH 29,775; MH 35,215 and a Research Scientist Award from the National Institute of Mental Health. I am grateful to Philipp Goedeking and David Symmes for comments on a previous draft of this chapter.

References

Baptista LF, Morton ML (1981) Interspecific song acquisition by a white-crowned sparrow. Auk 98:383–385

Baptista LF, Petrinovitch L (1984) Social interaction, sensitive phases, and the song template hypothesis in the white-crowned sparrow. Anim Behav 32:172–181

Bruner J (1983) Child's talk: learning how to use language. Norton, New York

Cheney DL, Seyfarth RM (1982) How vervet monkeys perceive their grunts: field playback experiments. Anim Behav 30:739–751

Chomsky N (1957) Syntactic structures. Mouton, The Hague

Cleveland J, Snowdon CT (1982) The complex vocal repertoire of the adult, cotton-top tamarin *(Saguinus oedipus oedipus)*. Tierpsychol 58:231–270

Dawson GA (1979) The use of time and space by the Panamanian tamarin, *Saguinus oedipus*. Folia Primatol 321:253–284

Dodd JJ (1972) Effects of social and vocal stimulation on infant babbling. Dev Psychol 7:80–83
Ferguson CA (1964) Baby talk in six languages. In: Gaumpertz J, Hymes D (eds) Ethnography of communication. Am Anthropol Suppl 66:103
Ferguson CA, Garnica OK (1977) Theories of phonological development. In: Snow C, Ferguson C (eds) Talking to children. Cambridge Univ Press, Cambridge
Kroodsma DE (1972) Variations in the songs of vesper sparrows in Oregon. Wilson Bull 84:173–178
Kroodsma DE (1973) Coexistence of Bewick's wrens and house wrens in Oregon. Auk 90:341–352
Kroodsma DE, Baylis JR (1982) Appendix: a world survey of evidence for vocal learning in birds. In: Kroodsma DE, Miller EH (eds) Song learning and its consequences. Acoustic communication in birds, vol 2. Academic Press, London New York, pp 25–49
Lemon RE (1975) How birds develop song dialects. Condor 77:385–406
Lenneberg EH (1969) On explaining language. Science 164:635–643
Marler P (1970a) A comparative approach to vocal learning: song development in white-crowned sparrows. J Comp Physiol Psychol 71:1–25
Marler P (1970b) Bird song and speech development: could there be parallels? Am Sci 58:663–669
Marler P, Peters S (1982) Subsong and plastic song: their role in the vocal learning process. In: Kroodsma DE, Miller EH (eds) Song learning and its consequences. Acoustic communication in birds, vol 2. Academic Press, London New York, pp 311–337
McConnell PB, Snowdon CT (1986) Vocal interactions between unfamiliar groups of captive, cotton-top tamarins. Behaviour 97:273–296
Mundinger P (1970) Vocal imitation and individual recognition of finch calls. Science 168:480–482
Newman JD, Symmes D (1982) Inheritance and experience in the acquisition of primate acoustic behavior. In: Snowdon CT, Brown CH, Petersen MR (eds) Primate communication. Academic Press, London New York, pp 259–278
Neyman PA (1977) Aspects of the ecology and social organization of free-ranging cotton-top tamarins *(Saguinus oedipus)* and the conservation status of the species. In: The biology and conservation of the Callitrichidae. Smithsonian Inst, Washington DC, pp 39–71
Oller DK, Wieman LA, Doyle WJ, Ross C (1975) Infant babbling and speech. Child Lang 1:1–11
Pepperberg IM (1986) Acquisition of anomalous communication systems: implications for studies in interspecies communication. In: Dolphin behavior and cognition. Erlbaum, New York
Pola YV, Snowdon CT (1975) The vocalizations of pygmy marmosets *(Cebuella pygmaea)*. Anim Behav 23:826–842
Seyfarth RM, Cheney DL (1986) Vocal development in vervet monkeys. Anim Behav 34:1640–1658
Snow C, Ferguson C (1977) Talking to children: language input and language acquisition. Cambridge Univ Press, Cambridge
Snowdon CT (1987) A naturalistic view of categorical perception. In: Harnad S (ed) Categorial perception. Cambridge Univ Press, Cambridge, pp 332–354
Snowdon CT (1988) A comparative approach to vocal communication. In: Leger DW (ed) Nebraska symposium on motivation, vol 35. Comparative approaches in modern psychology. Univ Nebraska Press, pp 145–199
Snowdon CT, Cleveland J (1980) Individual recognition of contact calls in pygmy marmosets. Anim Behav 28:717–727
Snowdon CT, Cleveland J (1984) "Conversations" among pygmy marmosets. Am J Primatol 7:15–20
Snowdon CT, Hodun A (1981) Acoustic adaptations in pygmy marmoset contact calls: location cues vary with distance between conspecifics. Behav Ecol Sociobiol 295–300
Snowdon CT, Pola YV (1978) Interspecific and intraspecific responses to synthesized pygmy marmoset vocalizations. Anim Behav 26:192–206
Thorpe WH (1958) The learning of song patterns by birds with especial reference to the song of the chaffinch *(Fringilla coelebs)*. Ibis 100:535–570
Todd GA, Palmer B (1968) Social reinformcement of infant babbling. Child Dev 39:591–596
Todt D (1975) Social learning of vocal patterns and modes of their application in grey parrots *(Psittacus erithacus)*. Z Tierpsychol 39:178–188
Todt D, Hultsch H, Heike D (1979) Conditions affecting song acquisition in nightingales *(Luscinia megarhynchos L.)*. Z Tierpsychol 51:23–35
Ziegler TE, Savage A, Scheffler G, Snowdon CT (1987) The endocrinology of puberty and reproductive functioning in female cotton-top tamarins *(Saguinus oedipus)* under varying social conditions. Biol Reprod 37:618–627

Conversational Vocal Exchanges in Squirrel Monkeys

DAVID SYMMES and MAXEEN BIBEN [1]

There is currently great interest in the reciprocal aspects of primate vocal communication. Words like "conversation" and "dialogue" are finding their way into published investigations of vocal behavior (Snowdon and Cleveland 1984; Maurus et al. 1985), no doubt because the essence of human vocal communication is conversation, and discovery of rudimentary conversational exchanges in lower forms would have considerable significance in both developmental psychology and neurology. The interest is certainly not new, but the emphasis may be. The view of nonhuman primate vocal communication as an interactive stream of signals which is modified by the presence of and the response by an audience appears frequently in this volume. It may be useful to consider some of the implications of the term conversation, especially when applied to the vocal behavior of monkeys and apes.

A definition of conversation useful to ethologists is difficult to obtain. Despite numerous efforts (see Laver 1976; McTear 1984 for reviews) the subject remains intractable; so inseparable from the human experience of information exchange through spoken language that animal models seem remote. Conversation is generally thought to be successful if both participants gain some information from it, but competence alone does not guarantee success (Tannen 1986).

Many examples come to mind of linguistic exchanges which are either ritualized or highly structured (such as orders from a superior) and fall short of even modest requirements for an exchange of ideas, intentions, needs, etc. In fact, without shared language competence, we have great difficulty deciding whether a conversation is taking place at all in our own species. Still, the obvious fact that conversational skills enrich and increase the efficiency of vocal communication is thought by all who study human language to be fundamental, and the origin of these skills is vigorously sought in early synchronous vocal exchanges between mother and child (Snow 1975).

An important constituent of human conversation which is likely to have a parallel in animal communication has been identified as indexical information (Abercrombie 1967). Indexical content refers to such matters as identity, attitude, and social history. Such information may include local accent, which has been a productive subject of research in bird song (Marler and Tamura 1964), and may become so in infrahuman primate research (Newman 1985). It is likely that vocal exchanges in the contexts of mating, care giving, and affiliation which have been described recently hold the greatest promise for locating the roots of human conversation. These sounds are of low inten-

1 Laboratory of Comparative Ethology, National Institute of Child Health and Human Development, Bethesda, MD 20892, USA

Todt/Goedeking/Symmes (Eds.)
Primate Vocal Communication

sity, as compared with territorial and alarm signals, and are intended for a small but significant (often closely related) audience.

If we are to propose that monkeys and apes engage in conversations, some critical examination of concepts is in order. Two properties usually attributed to human conversation are frequently offered as benchmarks against which all evidence from lower forms fall short. These are volitional control and spontaneity. It is our contention here that these concepts are poorly defined and cannot be used as bases for denying or affirming the existence of primitive conversation.

Volitional control is a flawed concept because no definition of choice is included. There is no doubt that vocal flexibility in monkeys and apes is restricted by the absence of direct neural pathways from cortex to brainstem motor neurons (Jürgens and Ploog 1970) and most evidence favors the view that vocal behavior is different from other motor performances in the ease with which it may be modified by instrumental conditioning procedures (Jürgens, this Vol., but also see Myers et al. 1965). These lines of reasoning suggest to us limitations, not impossibilities. The usual "push factor" explanation of the rich and varied sequencing of utterances in socially living monkeys requires one to propose a veritable kaleidoscope of rapidly changing motivational states, each producing the appropriate signal (see Scherer and Kappas, this Vol.). Yet observational data from several studies show that this vocal stream is not associated with behavioral responses which otherwise betray motivational changes, and in fact some evidence argues that the rate of vocalizing by monkeys can be quite decoupled from sensitive measures of physiological state (Levine et al. 1985). Indeed, the most common vocalizations found in resting, affiliating squirrel monkeys (Smith et al. 1982a) are motivationally neutral as assessed by brain stimulation techniques (Jürgens 1976). Although these difficulties with the evidence relating to volitional control of vocal behavior are substantial, we argue that the lack of an operational definition of choice is the crucial flaw[2]. We cannot assert the presence or absence of volitional control unless we know the behavioral repertoire available at the moment. This is possible only in highly structured situations in man, and impossible in the lower forms. We cannot say, for example, that the developing vervet monkey, which gives alarm calls to falling leaves as an inexperienced youth but acquires more reliable behaviors with maturity (Cheney and Seyfarth 1986), has or has not at any stage volitional control over its alarm calls.

Conversational spontaneity is another human behavior which is poorly defined. Sufficient controversy has been generated over chimpanzees trained to use manual signs and the claims that they converse in the absence of their trainers to make us all wary of spontaneity as a usable concept in animal behavior studies (Seidenberg and Petitto 1979). Yet the rich and continuous babble heard in both laboratory and field from groups of socially living monkeys prompts us to speculate that vocal behavior can be sustained in the absence of any reward or result other than the opportunity to hear others vocalize. Whether the babble is indeed spontaneous may remain in doubt, but we also suggest that it is particularly difficult to dismiss the notion of spontaneity in vocal behavior associated with play (Goedeking, this Vol.; Biben and Symmes 1986). The view that vocal behavior itself is a form of play at an early stage is traditional in

2 A similar argument has been presented by Menzel (1976).

child development research (Castner 1940), and it may not be unreasonable to see a phylogenetic parallel. The great difficulty in discussing spontaneity, however, is not unlike that found in arguments about volitional control. We cannot describe any observed behavior as spontaneous unless we are certain that all the internal states (primary needs) which might produce it are satisfied or at a neutral level. The claim for vocal spontaneity thus becomes a measure of the observer's ignorance rather than a measure of the performer's motivational framework.

If volitional control and spontaneity do not help our search for the roots of conversation, are there other ideas which may be more promising? We suggest three kinds of evidence which together can make a case for primitive, indexical conversations in monkeys and apes.

The first is turn taking, a concept which places the receptive aspect of vocal behavior on an equal basis with the productive aspect (cf. Snowdon, this Vol.). There has been a long-standing interest among ethologists in the detection of order in sequences of behavioral elements (Chatfield and Lemon 1970), and recent studies have applied these transitional probability techniques to vocal behavior (Snowdon and Cleveland 1984; Biben et al. 1986). It can be shown that in groups of monkeys the probability of an individual vocalizing depends on one or more prior calls. Such ordered vocalizing suggests that participants listen to and identify callers and share some mutually understood rules about vocal behavior which may resemble similar rules about food access, etc. Monkeys that chorus or vocalize repetitively with much overlap are probably not in that context taking turns, and the function of such behavior is more likely to lie in its effect outside the group than in its intragroup conversational potential (Sekulic 1982). The presence of an order of speaking which is nonrandom would appear to be a first requisite for conversation, but not in itself sufficient.

The next property of vocal exchanges we suggest is significant for demonstrating conversational properties is directionality of change in acoustic structure. The exchange must have a beginning and an end; conversations are not reversible and follow rules to avoid ambiguity (the implication that a universal grammar may apply to animals as well as humans is intentional; see Chomsky 1976). Repetition of stable elements such as that observed in duetting in gibbons (J.T. Marshall and E.R. Marshall 1976) cannot have this property and thus is not conversational. The precise distinction being offered here is progression within sequences which is independent of participant identity, as opposed to repeating a "learned part" even if acoustically different from the partners'. The fact that directional exchanges may be frequently repeated does not diminish their significance. Redundancy is not a conversational flaw.

The third critical property of vocal exchanges is alteration of vocal pattern (timing or structure) in a first speaker which is correlated with a vocal response from the listener (second speaker). This requirement is for evidence of information transfer, which is certainly not a new idea in communication. What is new in this argument is that the transfer must be bidirectional, as well as *within* the vocal framework. Freezing or fleeing responses of monkeys to alarm calls are clearly examples of information transfer but not evidence for conversation, even if the first caller desists when the alarm is taken. The most obvious example for a behavioral change within the linguistic realm is the question/answer paradigm. There is a distinction here from the earlier idea of order; B may vocalize following A (thereby establishing a transitional probability which may be significant), but A must in addition alter its vocal behavior (rate or form) as a result of B's response. Imitation of signals is not enough; mimicry is indeed a conversational flaw. There are to our knowledge virtually no studies in the literature on primate communication which have addressed this aspect of information transfer. Recent work by Maurus and collaborators (Maurus et al. 1985 and this Vol.) on functional boundaries within the vocal framework may be the first examples.

1 Experimental Studies

Recent findings from our laboratory at NIH have provided new evidence for conversational behavior in squirrel monkeys (Biben et al. 1986). We have confirmed earlier reports on the patterns of *chuck* calling among adult *Saimiri* during periods of quiet resting, affiliating, and care giving (Smith et al. 1982a). These patterns are robust and predictable. Even animals with little shared social history (captive bred and separately reared) will form affiliative networks and vocalize in the same way (see below) as do animals that were captured as adults and probably were members of a single troop. Despite some differences in social organization we have observed essentially similar patterns of affiliation and vocal behavior (Masataka and Biben 1987) in the Gothic-arch and Roman-arch species which we have studied (*Saimiri sciureus macrodon* and *S. boliviensis peruviensis* after Hershkovitz 1984).

The key feature of affiliative networks is individual preference. Pairs or less commonly triads of monkeys spend the majority of their resting time in proximity and/or huddling with each other. Some monkeys seem to distribute their preferences more widely (allowing us to define "popularity" operationally) while others are more mutually exclusive. The participants in such networks include all ages and sexes, but the most persistent (in terms of time budget) and the most faithful (in terms of partner preference) are adult females. Because of the conventions we follow (harem breeding) we have not had the opportunity to study such behavior in all-male groups. In harem groups we find that as a rule the adult male is the least active participant in affiliation. Juveniles often form their own affiliative groupings, and these appear to be based more on familiarity than sex.

Affiliative behavior as defined above implies a quiet, relaxed, unthreatened atmosphere in the monkey group. Familiarity with neighbors and the immediate environment, and (for captive animals) minimum human intervention are essential. Multiple perch spaces increase the time spent in affiliative behavior. Seldom are all members of a group huddling simultaneously, and squabbles over priorities or preferred locations can disrupt the relaxed atmosphere and inhibit affiliation even in animals not directly involved. Juvenile play normally does not disturb affiliative behavior, but can be disruptive when extremely active. For example, chasing play which includes jumping on huddle groups results in agonistic behaviors toward the juveniles and an increase in broadband calls.

Vocal behavior associated with affiliative behavior is extensive and reasonably homogeneous when considered at the level of classically defined call types (Newman 1985). Somewhere around 80% of the utterances are *chucks* (primarily tonal) and many of the rest are associated with momentary external disturbances or agonistic exchanges precipitated by space infringement, etc. Chuck calls are given by all adult females but some individuals are two or three times more vocal than others. We find that social usage of chucks is quite delayed in young squirrel monkeys (Biben and Symmes, unpublished observations) despite reports that such animals are capable of uttering them (Winter et al. 1973). There is no correlation between popularity and frequency of chuck calling, and no effect of distance between preferred partners up to the limits we can provide (about 4 m).

In a recent study we collected social and vocal behavioral data from a group of five adult female *S. s. macrodon* which were trapped as adults in the vicinity of Iquitos, Peru. It is unlikely that the individuals originated in common wild troops. Social behavior was recorded by two observers using microcomputers (Vic 20) with synchronized clocks, seated 1 m from the cage (1 x 3 x 1.8 m). Affiliative and agonistic behaviors were recorded by focal animal sampling, using huddling and proximity as measures of the former, and head grabbing for the latter. Vocalizations were recorded with cassette recorders (Sony TCM D5) fed by two omnidirectional microphones (Sony EM 170) placed less than 0.5 m from the animals. Sounds of vocal origin (bandwidth 0.5 to 16 kHz) were analyzed by a Voiceprint M700 sound spectrograph, using wide-band filtering. Intervals between vocalizations were measured with an accuracy of ± 50 ms from a strip chart made of the amplitude signal from tape. Speaker identities and call types were entered by hand on the strip chart from spectrograms. In all we recorded 1326 vocalizations in 260 min of observation, of which 83% were identifiable as chuck calls on the basis of criteria summarized by Newman (1985). The remainder included other recognizable call types, poorly recorded or interrupted calls. We created a data file in a larger computer (Digital 11-23) which contained for each chuck the identities of the speaker and the prior speaker, the time intervals surrounding the chuck, and eight parameters describing acoustic structure. More details of this method can be found in Biben et al. (1986).

We were then able to search this file for specific contexts (such as all chucks by A which followed a chuck by B) or specific interval characteristics. The subset derived from such searches could then be examined for acoustic properties or timing of repeating elements in the chuck. In Fig. 1 we have drawn the chuck call which exactly matches the average for each animal in eight characteristics. This illustrates the substantial individuality of the calls, which we have used in several studies to assist in identifying speakers (Smith et al. 1982b). Individuality is encoded in frequency and duration parameters, including the occurrence, peak frequency, shape, and duration of the flag, peak frequency, and slope of the masts, and interval between the masts. Of greater interest in the search for vocal details suggestive of conversations is the use of these data to reveal changes correlated with the position of a chuck in a sequence, and the generality of such changes across individuals. We have concluded that there is evidence for conversational usage, which may be summarized as follows.

1. Evidence of turn taking. In a sequence of chucks, the identity of successive vocalizers is not random. We carried out an information theory analysis, following the methods and nomenclature of Chatfield and Lemon (1970). The results strongly suggest an order of speaking or turn taking that depends on recognition of individuals' chucks. Transitional probabilities for chuck sequences indicate significant gains in information given knowledge of the prior caller (H2 = 2.249) or prior two callers (H3 = 2.104;

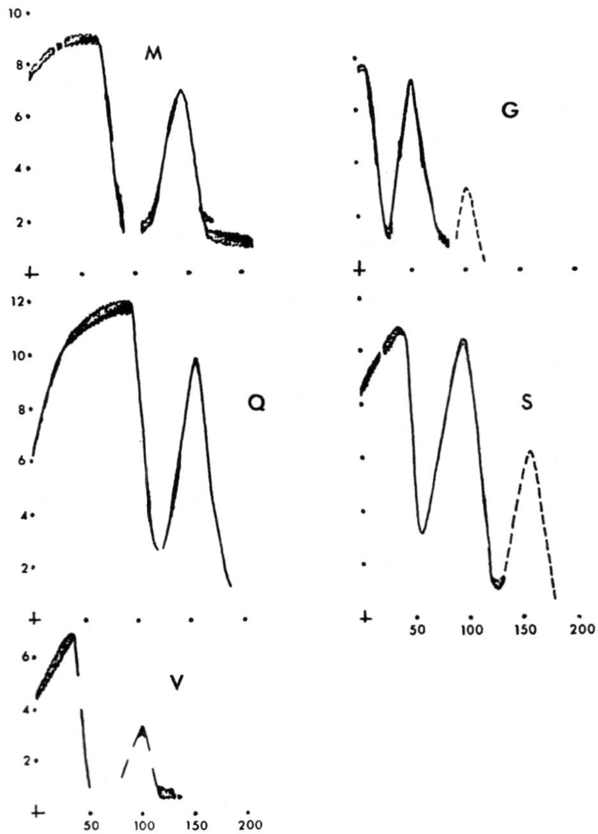

Fig. 1. Redrawn sound spectrograms of a typical chuck from each of five female Gothic-arch squirrel monkeys. *Saimiri sciureus macrodon*, identified by a single letter code. Nomenclature for chuck elements is from Newman et al. (1983): *Flags* are drawn to show the most common shape for each animal (none for subject *G*). Only *G* and *S* used three mast chucks. Ordinate is in kHz, abscissa in ms

$p < 0.001$ in both cases). Affiliative preference and calling sequences were correlated in that females were more likely to call following the chucks of more preferred partners (an independent confirmation of prior findings reported in Smith et al. 1982a).

2. Temporal spacing of chucks in a sequence is nonrandom. In Gothic-arch squirrel monkeys, latency and affiliative preference are positively correlated in a monotonic relationship (Fig. 2). In Roman-arch animals, latency is more formalized such that animals follow a chuck from a preferred partner with their own within 0.5 s or keep silent. Latencies thus are distributed bimodally (Masataka and Biben 1987). At this point we may speak of a chuck that promptly follows another as a "response" to that chuck.

3. The structure of an individual's chuck varies with the position in the sequence. Using temporal dependency as the criterion, comparison was made between an individual's chucks occurring in prompt response to another's chucks, with those preceded by > 5 s of vocal silence. The latter may be termed "questions" and the former "answers" (Table 1). Chucks serving as questions differ from those serving as answers in the peak frequency attained at several points in the course of the call, including the flag, first mast, and second mast (if present). Answer chucks were about 300 Hz higher than question chucks, a frequency difference which we estimate to be well within the limits

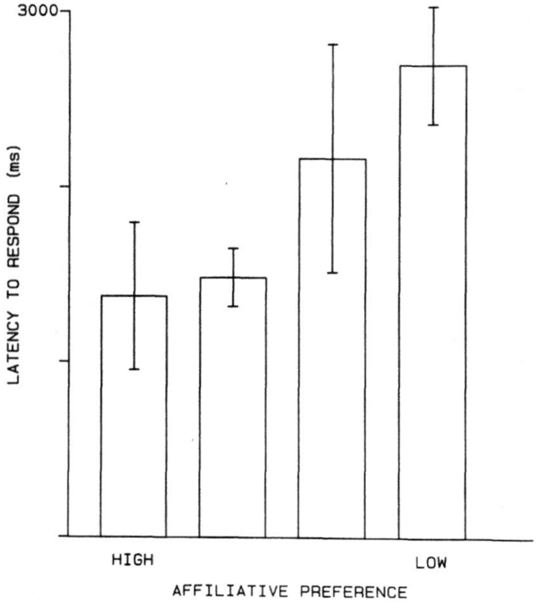

Fig. 2. Relationship between affiliative preference and latency to respond (with a chuck) to the chuck of another individual. Affiliative preference is based on the amount of time each of five females spent huddling with the other four. Females responded with the shortest latency to the chucks of individuals with whom they affiliated most often

Table 1. Frequency measures compared for "question" and "answer" chucks from five female squirrel monkeys, *S. sciureus macrodon*. Analysis of variance between conditions was significant for each measure. Analysis of variance within subjects also revealed clear individuality in the chuck. (PFF1 = peak frequency in Hz attained within the first flag, if present; PFM1 = peak frequency at top of first mast; PFM2 = peak frequency at top of second mast, if present)

Subjects			G	M	S	Q	V	Group mean
PFF1:	Questions	N	21	30	43	57	85	9438
		$\bar{X}$	8232	9040	11111	11896	6910	
		SD	1334.8	340.2	635.0	473.1	694.3	
PFF1:	Answers	N	28	69	63	59	62	9839
		$\bar{X}$	9535	9277	11125	12123	7136	
		SD	1554.8	448.6	604.7	409.0	735.0	
	$F_{between\,[1,507]} = 20.57^{***}$			$F_{within\,[4,507]} = 956.44^{***}$				
PFM1:	Questions	N	62	30	51	57	85	9244
		$\bar{X}$	7737	8924	10779	11871	6910	
		SD	1708.4	333.6	1314.0	488.9	694.9	
PFM1:	Answers	N	90	70	76	60	62	9513
		$\bar{X}$	8233	9171	10930	12096	7133	
		SD	2162.7	447.1	1075.8	408.1	729.3	
	$F_{between\,[1,633]} = 8.43^{**}$			$F_{within\,[4,633]} = 388.0^{***}$				
PFM2:	Questions	N	59	15	50	48	49	7418
		$\bar{X}$	7110	6962	10270	9688	3059	
		SD	1161.1	892.4	1170.7	1200.8	1549.0	
PFM2:	Answers	N	86	49	75	53	38	7847
		$\bar{X}$	7623	7272	10404	10400	3534	
		SD	1303.7	965.1	1049.8	706.4	1450.1	
	$F_{between\,[1,512]} = 15.87^{***}$			$F_{within\,[4,512]} = 550.77^{***}$				

** $p < 0.01$, *** $p < 0.001$.

of *Saimiri* discrimination, based on indirect evidence (see Biben et al. 1986). Such differences overlaid the clear individuality present in frequency measures of chucks, suggesting that females must have some familiarity with the speaker to know whether any chuck was her high chuck (answer) or low chuck (question). It is significant that the frequency difference between question and answer chucks was present and in the same direction for all animals. The consistency of these data encourage us to describe this communication system as stable, based on knowledge of individual characteristics (indexical), and operating under formalized rules. In contrast to frequency parameters, duration measures (flag duration and intermast interval) did not vary with the position in sequence. It is not surprising that these measures do not carry conversational information since flag occurrence and number of masts are highly individual characteristics, with some animals routinely using no flag and/or only one mast.

4. Structural variability in answers correlates with subsequent vocal behavior of the questioner. Again using the criterion of prompt response, we compared answers that were themselves promptly followed by a chuck from the questioner with those that terminated the sequence (i.e., were followed by silence exceeding the median response times of all animals). In simple terms, we asked whether a persistent questioner would desist if a satisfactory (high) answer were obtained. Terminating chucks had higher frequency measures for the first flag and first mast, when compared to chucks that continued the sequence. The effect was not detectable on later (second mast) peak frequencies (Table 2).

Table 2. Frequency measures compared for "terminating" and "continuing" answer chucks. Subjects and measures are the same as in Table 1

Subjects			G	M	S	Q	V	Group mean
PFF1:	Terminating	N	5	26	23	25	25	9928
		$\bar{x}$	9758	9284	11202	12230	7164	
		SD	559.1	398.6	538.3	323.3	527.2	
PFF1:	Continuing	N	5	19	14	12	8	9626
		$\bar{x}$	9257	9108	11172	11790	6804	
		SD	1441.8	344.9	734.1	422.3	393.7	
	$F_{between\ [1,152]} = 8.57^{**}$			$F_{within\ [4,152]} = 484.85^{***}$				
PFM1:	Terminating	N	29	27	27	25	25	9579
		$\bar{x}$	8265	9133	11117	12216	7164	
		SD	2055.7	386.1	550.9	321.1	527.2	
PFM1:	Continuing	N	15	19	18	12	8	9200
		$\bar{x}$	7494	9015	10928	11761	6804	
		SD	2307.5	352.0	993.4	393.7	393.7	
	$F_{between\ [1,195]} = 5.11^{*}$			$F_{within\ [4,195]} = 135.84^{***}$				
PFM2:	Terminating	N	28	16	27	24	16	7800
		$\bar{x}$	7593	7357	10285	10236	3530	
		SD	1179.2	680.7	1264.2	703.1	909.0	
PFM2:	Continuing	N	15	14	18	10	5	7696
		$\bar{x}$	7490	7512	10478	10311	2690	
		SD	1197.5	676.0	1164.7	536.4	833.3	
	$F_{between\ [1,163]} = 0.01^{a}$			$F_{within\ [4,163]} = 219.51^{***}$				

* $p < 0.05$, ** $p < 0.01$, *** $p < 0.001$.
[a] Not significant.

These data provide evidence that questions and answers may be distinguished structurally and that formal rules (syntax) govern the employment of these variants. An inference from these data is that animals are seeking information by questioning and receiving information from answers.

2 Discussion

One goal in the study of nonhuman primate communication is the discovery of continuities between such natural systems and human language. A handicap to this pursuit has been, in our view, a preoccupation with theoretical orientations which polarizes professional scientists and laymen and produces a tendency to draw conclusions which go beyond available evidence. The opposing views are called by several names, but most often the nativist and mentalist approaches.

Our approach is pragmatic and is founded upon use of a rich data base with good temporal resolution and attribution of all vocalizations. In *Saimiri*, chuck exchanges were a good choice because they are produced abundantly, are correlated with affiliative contexts, and are individually recognizable and tonal in nature. Similar vocalizations are probably present in many other gregarious primates. Multiple measurements and statistical comparisons did in this case, and will no doubt in others, reveal a good deal about the patterns of information transfer.

We have described behavior in squirrel monkeys which appears to us to resemble human conversation, and we have used this term in its conventional meaning (which is albeit somewhat elusive) rather than in any special, reduced sense implied by setting it in quotation marks (Snowdon and Cleveland 1984). The significance of finding primitive conversations in monkeys should lie in encouragement of similar lines of research in several laboratories rather than in refutation or confirmation of some philosophical orientation. It is not necessary to impute awareness or conscious intention to squirrel monkeys to accept the present evidence. A narrow view of monkeys seeking and repeatedly providing indexical information in a primitive conversational format and thereby enhancing fitness through sociality is reasonable and conservative. It may be that chuck exchanges in squirrel monkeys resemble anal sniffing in canids more than they resemble human language, but we would also suggest that they share some rather interesting properties with human conversation. These properties include recognition by voice, turn taking, syntactical conventions or rules, bidirectional information exchange, and limitation of the interaction to the vocal level.

References

Abercrombie D (1967) Elements of general phonetics. Edinburgh Univ Press, Edinburgh

Biben M, Symmes D (1986) Play vocalizations of squirrel monkeys *(Saimiri sciureus)*. Folia Primatol 46:173–182

Biben M, Symmes D, Masataka N (1986) Temporal and structural analysis of vocal exchanges in squirrel monkeys *(Saimiri sciureus)*. Behaviour 98:259–273

Castner BM (1940) Language development. In: Gesell A et al. (eds) The first five years of life. Harpers, New York

Chatfield C, Lemon RE (1970) Analysing sequences of behavioural events. J Theor Biol 29:427–445

Cheney D, Seyfarth R (1986) Vocal development in vervet monkeys. Anim Behav 34:1640–1658

Chomsky N (1976) On the nature of language. Ann NY Acad Sci 280:46–58

Hershkovitz P (1984) Taxonomy of squirrel monkeys genus *Saimiri* (Cebidae, Platyrrhini). Am J Primatol 6:257–312

Jürgens U (1976) Reinforcing concomitants of electrically elicited vocalizations. Exp Brain Res 25:203–214

Jürgens U, Ploog D (1970) Cerebral representation of vocalization in the squirrel monkey. Exp Brain Res 10:532–554

Laver J (1976) Language and nonverbal communication. In: Carterette EC, Friedman MP (eds) Handbook of perception, vol VII. Academic Press, London New York

Levine S, Johnson DF, Gonzales CA (1985) Behavioral and hormonal responses to separation in infant rhesus monkeys and mothers. Behav Neurosci 99:399–410

Marler P, Tamura M (1964) Culturally transmitted patterns of vocal behavior in sparrows. Science 146:1483–1486

Marshall JT Jr, Marshall ER (1976) Gibbons and their territorial songs. Science 193:235–237

Masataka N, Biben M (1987) Temporal rules regulating vocal exchanges of squirrel monkeys. Behaviour 101:311–319

Maurus M, Kuehlmorgen E, Wiesner E, Barclay D, Streit KM (1985) 'Dialogues' between squirrel monkeys. Lang Commun 5:185–191

McTear MF (1984) Structure and process in childrens' conversational development. In: Kuczaj SA (ed) Discourse development. Springer, Berlin Heidelberg New York Tokyo

Menzel E (1976) Discussion of paper by Mayers RE. Ann NY Acad Sci 280:758–760

Myers SA, Horel JA, Pennypacker HS (1965) Operant control of vocal behavior in the monkey *Cebus albifrons*. Psychonom Sci 3:389–390

Newman JD (1985) Squirrel monkey communication. In: Rosenblum LA, Coe CL (eds) Handbook of squirrel monkey research. Plenum, New York London

Newman JD, Smith HJ, Talmage-Riggs G (1983) Structural variability in primate vocalizations and its functional significance: an analysis of squirrel monkey chuck calls. Folia Primatol 40:114–124

Seidenberg MS, Petitto LA (1979) Signing behavior in apes: a critical review. Cognition 7:177–215

Sekulic R (1982) The function of howling in red howler monkeys *(Alouatta seniculus)*. Behaviour 81:38–54

Smith HJ, Newman JD, Symmes D (1982a) Vocal concomitants of affiliative behavior in squirrel monkeys. In: Snowdon CT, Brown CH, Petersen MR (eds) Primate communication. Cambridge Univ Press, Cambridge

Smith HJ, Newman JD, Hoffman HJ, Fetterly K (1982b) Statistical discrimination among vocalizations of individual squirrel monkeys *(Saimiri sciureus)*. Folia Primatol 37:267–279

Snow CE (1975) The development of conversation between mothers and babies. J Child Lang 4:1–22

Snowdon CT, Cleveland J (1984) "Conversations" among pygmy marmosets. Am J Primatol 7:15–20

Tannen D (1986) That's not what I meant! Morrow, New York

Winter P, Handley P, Ploog D, Schott D (1973) Ontogeny of squirrel monkey calls under normal conditions and under acoustic isolation. Behaviour 47:230–239

Vocal Play Behavior in Cotton-Top Tamarins

PHILIPP GOEDEKING[1]

Categorizing an animal's vocal repertoire into classificatory units is usually one of the first undertakings when studying its vocal behavior. A persistent problem in evaluating such catalogues in primates is the enormous structural variability of many primate vocalizations. Repertoire descriptions of the same species often vary dramatically among authors (e.g. *Saimiri sciureus*: Winter et al. 1966; Schott 1975; Jürgens 1979; Newman 1985).

Categorizing by a finite set of discrete clusters may be inadequate to describe the variety of vocal patterns of many primate species. Many classificatory schemes even appear to be restatements of initial a priori assumptions. Significant differences between preconceived classes, if not biologically relevant, may only be evidence of the researcher's adroitness in applying classificatory rules to data. Thus, even consistent classification provides only marginal information if it is not demonstrated that the classification rules reflect functional properties.

It is suggested here to consider a vocalization as (1) a configuration of vocal features rather than a "gestalt", and (2) to investigate the covariation between or among vocal features and with ongoing behavior. Such an approach is not biased by vague assumptions on the graded or discrete nature of calls. Beyond that, studying vocal features may illustrate aspects of the natural history of vocal shaping.

Acoustic properties of vocal patterns cannot be assumed to be selectively neutral. The level at which selection occurs is not necessarily the vocalization as a whole, but rather its acoustic properties. Graded factors, like sound transmission characteristics of the habitat, are known to exert selective pressures on the acoustic properties of vocal signals. Selection favoring uniform configurations of vocal features, hence "types", has presumably occurred due to the ease of perception of such patterns (Symmes 1981).

U. Scherer et al. (1980; K. Scherer 1985; K. Scherer and Kappas, this Vol.) suggested calling emotional factors which influence vocal shape "push factors" in contrast to external "pull factors". The former were defined (U. Scherer et al. 1985, p. 209) "... as those changes in the states of the internal support and action subsystems which affect the production of vocalization in an essentially nondirective manner". On the other hand, pull factors are "... norms or expectations imposed by the external physical or social environment which require the production of specific acoustic features ... to achieve a particular effect". Remembering Lack's concept of ultimate and proximate

1 Institut für Verhaltensbiologie, Fachbereich Biologie, Freie Universität Berlin, Haderslebener Straße 9, 1000 Berlin 41, FRG

Todt/Goedeking/Symmes (Eds.)
Primate Vocal Communication

factors ("Ultimate factors are concerned with survival value, proximate factors with adaptations in physiology and behavior", Lack 1954), external pull factors can be recognized as ultimate factors, since they determine the benefits and risks of vocal behavior, but not its physical realization. Internal push factors can be identified as the respective proximate mechanisms. The key to this developmental perspective is making clear that selection can operate only via external pull factors. Thus, vocal qualities are likely to reflect both the ultimate needs and proximate repercussions involved in vocal signaling.

Cotton-top tamarins *(Saguinus oedipus oedipus)* exhibit a rich example of increased use of particular vocal qualities during playful wrestling. Pitch instability, pitch range, and pitch mode each covary with ongoing wrestle play, and will be considered in detail in this chapter.

1 Methods

The observed group consisted of six individuals: two parents about 5 years of age, and two pairs of twins (one pair 2 years and the other 2 months of age). The animals were housed in a spacious indoor cage (ca. 400 × 250 × 210 cm), located in a room isolated from other animals. The walls of the room were covered with foam material to attenuate sound reflection. The interior of the cage was enriched with ropes, branches, and a variety of toys. Simultaneous audio- and videorecordings were made using a Revox B77 tape machine with a Sennheiser TU416 microphone and a National b/w-videosystem. The recording position was approximately 1.5 m from the front side of the cage. The animals were well habituated to this setting. During the observations, a computer-generated time code (ROCKWELL AIM 65) was written on the second channel of the tape and the sound channel of the videorecorder. By reading the time code each vocalization could be related with its associated behavior. The videorecordings were visually inspected in steps of 2.5-s intervals on a videomonitor. A character code for the behavior observed during each 2.5-s interval was typed into the computer. Thus, a character sequence describing behavioral data was linked to the corresponding acoustic data.

The vocalizations were directly fed from the tape recorder into a minicomputer (Data General Eclipse AP 130). Frames of 256 consecutive points were Fast Fourier transformed (FFT) to build up the amplitude spectrum (bandwidth = 10 kHz, time resolution = 12.8 ms, frequency resolution = 78.125 Hz). A Fortran program first generated a spectrographic representation and then computed F_0 for each spectrum. The onset as well as the end of a vocalization was manually marked on the graphic monitor by moving a crosshair across the screen. The stepwise sequence of F_0-indicators in between these markers formed the pitch contour of that particular vocalization, which was stored in a file for further processing. In total, 3066 vocalizations were quantitatively analyzed for this study.

Pitch instability describes deviations of the pitch time course from a very smooth line. The procedure for quantifying pitch contour instability is described in detail elsewhere (Goedeking and Immelmann 1986) and will be developed here only intuitively: The computer moves 128 "fingers" along the time axis across the pitch contour of the inspected vocalization. The repetitive touching of the pitch contour to certain fingers

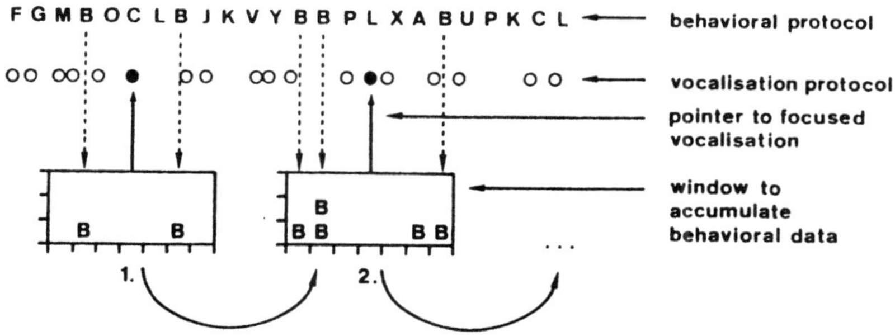

Fig. 1. Evaluation of a PPP: a "window" (in this graph reduced to seven time intervals) is overlaid to a specified vocalization (indicated by a *block dot*). The occurrence of a "B" symbol in the behavioral data protocol is stored in the window. Then the window shifts to the next specified vocalization and all "B"s were again accumulated on the previous profile and so on (Goedeking and Immelmann 1986)

(measured as "k") is in direct proportion to the degree of the highs and lows in that contour. Pitch range is computed as the difference between the highest and the lowest point within a single vocalization. Pitch mode shall be defined as the pitch value which occurs most often in a given vocalization (given the aforementioned limited frequency resolution of 78.125 Hz).

For analysis of covariation between the occurrence of vocalizations with particular structural features (such as a specific degree of instability, mode, or range in pitch) a windowing technique was developed, whereby a 62.5-s window overlaid any specified vocalization to accumulate the occurrence of wrestle-play symbols for the period 30 s before and after the vocalization in question (Fig. 1). The window was subdivided into 25 intervals of 2.5 s each, with one central interval covering the moment of the focused vocalization. Whenever in a 2.5-s interval the wrestle-play symbol "B" occurred, it was copied onto the respective interval of the window. For the next specified vocalization all intervals with B were added to the previous profile and so on. The combined profile (called here a "Pre-Post-Profile", PPP) represents the rate of wrestle play found 30 s before and after vocalizations selected for some particular structural quality.

For purpose of interpretation, a PPP of, for example, the most instably pitched vocalizations has to be compared with the PPP of the most stably pitched calls. Then, the differences in the highs and lows of the compared PPPs can be attributed to respective contextual differences. Arranging one PPP into relation to another results in a sequence of quotients, the ups and downs of which indicate the differences of interest. A PPP being related to another one shall be referred to here as a related pre-post-profile, RPPP.

The significance of observed RPPPs was evaluated by generating a chance distribution by random selection of vocalizations. "Random-RPPPs" were accumulated in the same manner as described above. If, for a given interval, the values in the original RPPP exceed ranges of 50 random-RPPPs (as determined by Student's t-distribution), the former is said to significantly differ from the latter.

2 Results

Of all calls in this study 78% showed no pitch instability at all (k = 0), 10 % had moderate (0 > k ⩽ 2), 12% high scores (2 < k) of pitch instability. As the "pitch-instability-RPPP" in Fig. 2 shows, the rate of wrestling reaches a significant maximum only a few seconds after those vocalizations which have a high degree of pitch instability. The values rapidly decrease thereafter and approach a significant minimum about 20 s later. In other words: a few seconds after instably pitched vocalizations a significantly increased rate of wrestle play occurs, and then about 20 s later a significantly reduced rate.

The graph in Fig. 3 shows the distribution of pitch values found in all spectra in all of the 3066 vocalizations. This information is needed to determine ranges of pitch mode appropriate to be correlated with play.

Obviously, pitch is preferably intonated between 1400 and 3500 Hz with a strikingly high maximum at about 2400 Hz. A very low pitched call, therefore, should have a pitch mode in the lower half of this peak.

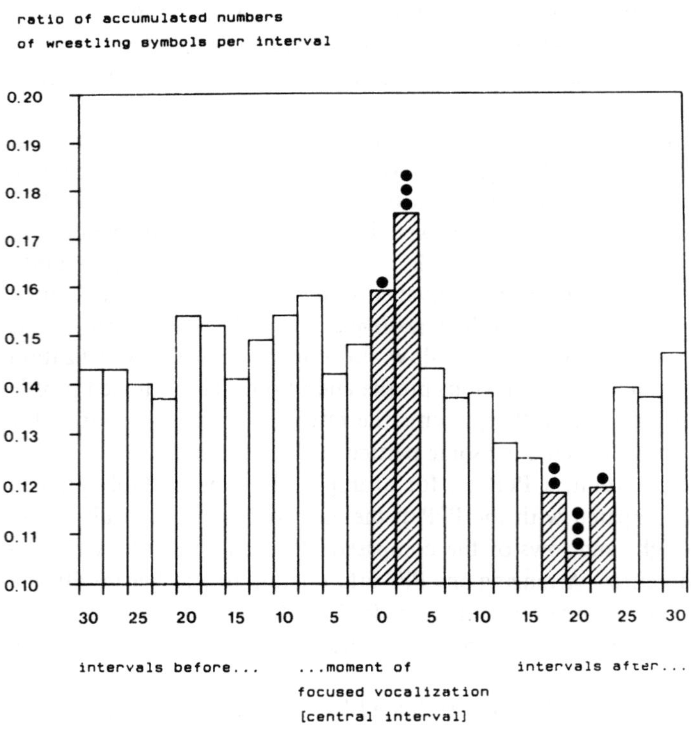

Fig. 2. Pitch-instability RPPP [PPP of high pitch instability (k = 2) over PPP of no pitch instability (k = 0)]. •: $p \leqslant 0.05$; ••: $p \leqslant 0.01$; •••: $p \leqslant 0.001$

Fig. 3. The distribution of F_0 values found in all spectra in all 3066 vocalizations

Fig. 4. Low-pitch RPPP [PPP of low-pitched calls (mode: 1406–2265 Hz) over PPP of high-pitched calls (mode: 2265–3516 Hz)]. •: $p \leqslant 0.05$; ••: $p \leqslant 0.01$; •••: $p \leqslant 0.001$

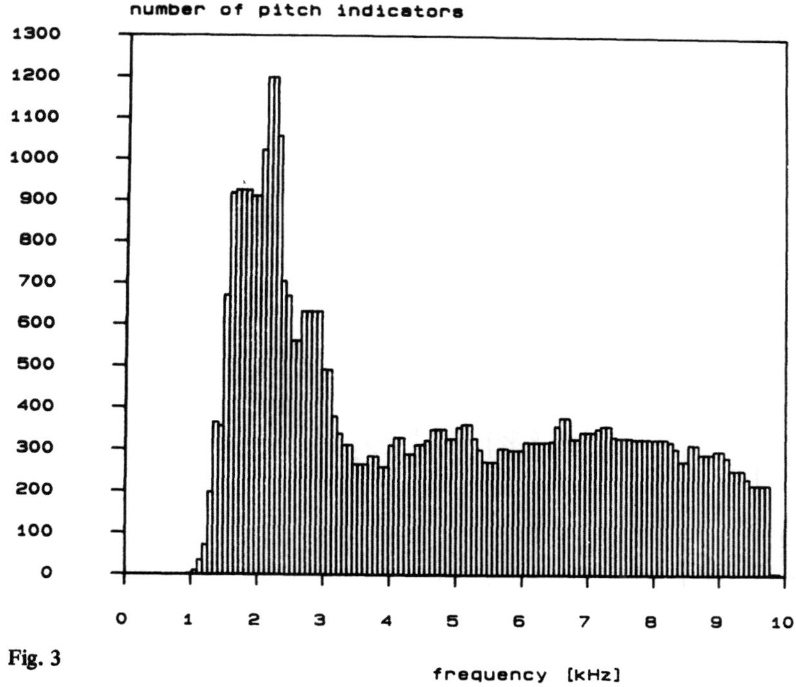
number of pitch indicators
1300
1200
1100
1000
900
800
700
600
500
400
300
200
100
0
0 1 2 3 4 5 6 7 8 9 10
frequency [kHz]

Fig. 3

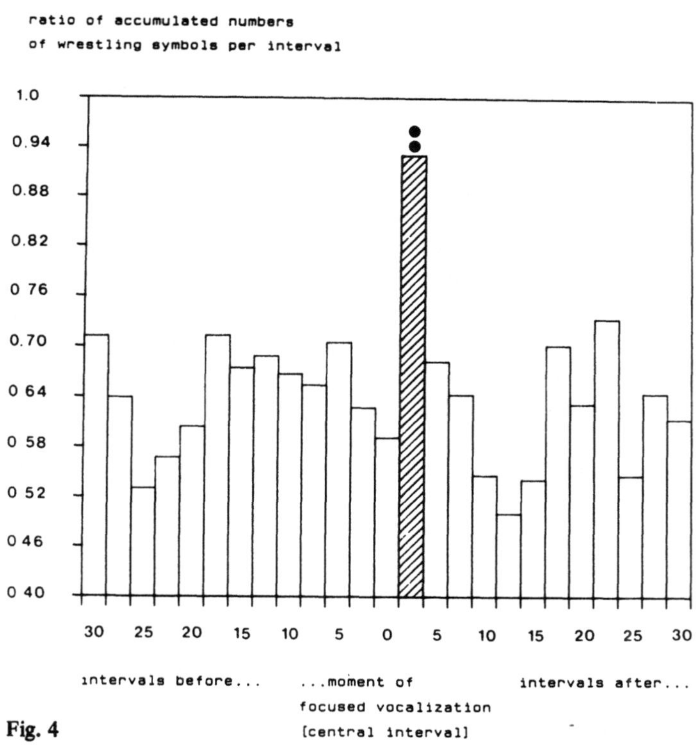
ratio of accumulated numbers
of wrestling symbols per interval
1.0
0.94
0.88
0.82
0 76
0.70
0 64
0 58
0 52
0 46
0 40
30 25 20 15 10 5 0 5 10 15 20 25 30
intervals before...
...moment of
focused vocalization
[central interval]
intervals after...

Fig. 4

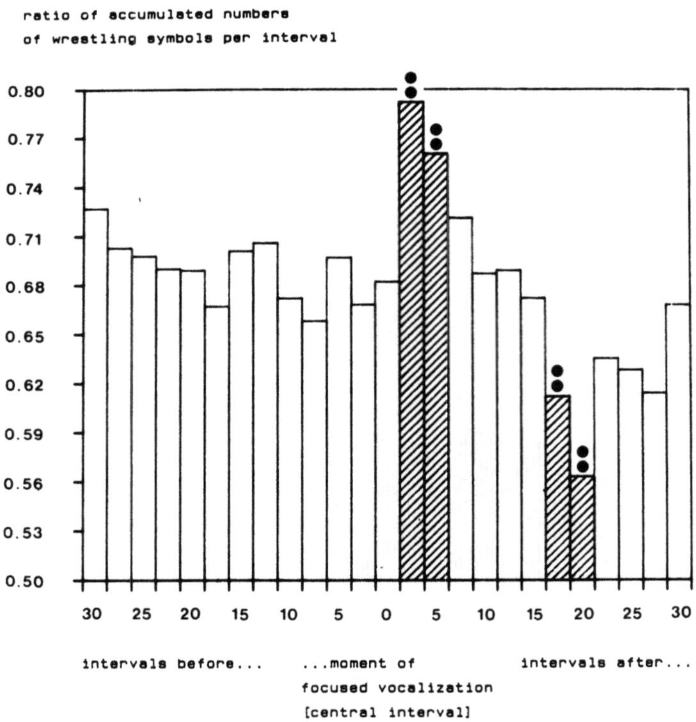

Fig. 5. Narrow-range RPPP [PPP of narrow ranging calls (range: 0–781 Hz) over PPP of wide ranging calls (mode: more than 781 Hz)]. •: $p \leqslant 0.05$; ••: $p \leqslant 0.01$; •••: $p \leqslant 0.001$

The "low-pitch-RPPP" (Fig. 4) resembles that of the instably pitched vocalizations to a high degree: the maximum rate of wrestle play occurs within the interval immediately following the central interval covering the focused vocalization. As in the case of pitch instability, a wrestle play minimum is found about 20 s later. Thus, there is a significant interdependence between wrestle play and very low pitched vocalizations.

To determine biologically appropriate magnitudes of pitch range, reference is given to the pitch distribution shown in Fig. 3. A vocalization with a narrow pitch range shall be defined as occurring when the highest and the lowest pitched part of a vocalization are less than 780 Hz apart from each other (that is, as narrow as the central portion of the peak in Fig. 3).

Again, this "narrow-range-RPPP" (Fig. 5) shows a shape similar to that in Figs. 2 and 4.

3 Discussion

At least three structural features in cotton-top tamarin vocalizations provide examples of contextual concentrations of particular vocal qualities: Where the vocalizations are characterized by a high degree of instability in the contour of the fundamental frequency, by a very restricted pitch range, and/or by a very low pitched mode, one finds an increased rate of wrestle play.

Instable pitch contours have been largely investigated in studies on human voice. Instable pitch in human voicing has been found to express affective states of high arousal (K. Scherer 1985). Consistent with studies on pathological alterations of voicing in human beings (Davis 1979), pitch instability is assumed to result from a drop in fine motor control on the larynx in situations of high emotional excitement.

Caution is needed in comparing results derived from human voice pitch instability with the same parameter measured in the small arboreal New World monkey. A statement on what is stable or instable pitch requires a species-specific determination on what is normal in this respect. Pitch of tonal calls in many arboreal New World monkey species is in the range of several kHz, whereas the F_0 of human speech is in the magnitude of a few 100 Hz at the most. The vocal folds in these primate species are short, requiring tense vocal cords and high subglottal pressures to produce high pitched sounds. A greater extent of pitch instability is expected in the high pitched calls of New World monkeys as compared to the low pitched human voice.

As this report has shown, most vocalizations of cotton-top tamarins show no evidence of pitch instability. This finding contrasts with Lieberman's (1968, 1969) argument that subhuman primates in general are characterized by an instable pitch contour (see also: Seyfarth and Cheney 1984).

A few examples of context-related occurrence of instably pitched calls have been shown in other subhuman primates. Stevenson and Poole (1982) presented a spectrogram of a considerably irregular pitched call of the common marmoset *(Callithrix j. jacchus)*, which they reported typically appears during play bouts. Jürgens (1979) concluded in a study on the aversive or pleasurable concomitants of squirrel monkey vocalizations, that "... a comparison of the long duration calls [within the "chirping-peeping-squeeling" group] ... reveals that calls with a relatively smooth frequency course ... are less 'aversive' than calls with an irregular frequency course". In a study on the vocal repertoire of the cotton-top tamarin, Cleveland and Snowdon (1982) correlated some vocalizations with play contexts. As inspection of these sonagrams reveals, these vocalizations are among the most instably pitched in the list included in that work.

In summary, these findings corroborate the conclusion that, as in human speech, a high degree of pitch instability in nonhuman phonation corresponds to a considerably high level of emotional involvement. It is likely that in the case of cotton-top tamarins it is not primarily playful wrestling itself, but rather some kind of emotional excitement involved in play that causes the appearance of instable pitch.

Vocalizations with a very limited pitch range and restricted to a sharp-edged low frequency bandwidth are highly qualified for long-range transmission with only minor degradation. The sharp peak of preferred pitch matches that frequency bandwidth known to attenuate sound less than other frequencies in some horizontal strata of the tropical rain forest (Marten and Marler 1977). Additionally, these acoustic features provide many localizable cues, thus potentially attracting predators. Hershkovitz (1977) notes that "tamarins live under constant threat of attack during the day by raptorial birds, felids, mustelids, and probably snakes". Moynihan (1970) reports a tyra *(Eira barbata)* seen with a dead tamarin in its mouth. Though it is not clear to what extent vocalizations contribute to this "constant threat of attack", the selective advantages of such vocalizing during play should exceed the risk of enhanced detectability by predators.

There are at least two possible explanations for this kind of vocal behavior, one of which seeks in the ultimate, the other in the proximate, domain.

Biben and Symmes (1986), in a study on squirrel monkey play vocalizations, suggested the possibility that due to the long-range transmission quality of play vocalizations in this species, these calls might be directed to other group members rather than the actual play partner. The same may apply here. Unfortunately, no laboratory or field observations are available so far to support or disprove this hypothesis.

The alternative hypothesis is rather speculative, but takes into account many of the concepts characterizing social play as "pleasurable", "exaggerated", and "exciting": Whatever the ultimate function of play may be, its pleasurable or exciting concomitants might serve as proximate mechanisms to ensure its appearance. This notion receives support from findings in rats, where opiates were found to increase play, whereas the respective antagonists reduce it (Panksepp et al. 1980; Beatty et al. 1982). As many other forms of expressive behavior (e.g., gestures, facial expression, piloerection), the rate of vocalization as well as the acoustic structure are likely to be affected by such positively reinforcing emotional processes. Due to the essentially nondirective manner of affective vocal shaping, the respective acoustic outcome has not necessarily to be adaptive in terms of fitting social or environmental needs ("pull factors"). Production as well as hearing of play vocalizations, the structure of which is mainly effected by physiological manifestations of emotional processes, may even reinforce the positively rewarding impact of ongoing play. Ultimate benefits of such vocal behavior may well lie in its being self-directed. The risk of vocalizing could be the price for the fun connected with it, and for individuals of those species which play quietly, this price may be too high.

Acknowledgments. I am grateful to M. Biben, K. Immelmann, D. Symmes, D. Todt, and many others for help and/or comments on earlier versions of the manuscript. Research reported here was in part supported by a grant from the DFG (IM 1/16-1).

References

Beatty W, Dodge AM, Dodge LJ, White K, Panksepp J (1982) Psychomotor stimulants, social deprivation and play in juvenile rats. Pharmacol Biochem Behav 16:417–422

Biben M, Symmes D (1986) Play vocalizations of squirrel monkeys *(Saimiri sciureus)*. Folia Primatol 46:173–182

Cleveland J, Snowdon CT (1982) The complex vocal repertoire of the adult cotton-top tamarin *(Saguinus oe. oedipus)*. Z Tierpsychol 58:231–270

Davis S (1979) Acoustic characteristics of normal and pathological voices. In: Lass NJ (ed) Speech and language. Academic Press, London New York, pp 273–338

Goedeking P, Immelmann K (1986) Vocal cues in cotton-top tamarin play vocalizations. Ethology 74:219–224

Hershkovitz P (1977) Living New World monkeys *(Platyyhini)*. Univ Chicago Press, Chicago

Jürgens U (1979) Vocalizations as an emotional indicator. A neuroethological study in the squirrel monkey. Behaviour 69:88–117

Lack D (1954) The natural regulation of animal numbers. Oxford Univ Press, Oxford

Lieberman P (1968) Primate vocalizations and human linguistic ability. J Acoust Soc Am 44:1574–1584

Lieberman P (1969) On the acoustic analysis of primate vocalizations. Behav Res Methods Instrum 1:169–174

Marten K, Marler P (1977) Sound transmission and its significance for animal vocalization. I Temperate habitats. Behav Ecol Sociobiol 2:271–290

Moynihan M (1970) Some behavioral patterns of platyrrhine monkeys II: *Saguinus geoffroyi* and some other tamarins. Smithsonian Contrib Zool 28:1–77

Newman JD (1985) Squirrel monkey communication. In: Rosenblum LA, Lee CK (eds) Handbook of squirrel monkey research. Plenum Press, New York, pp 99–126

Panksepp J, Herman BH, Vilberg T, Bishop P, DeEskinazi FG (1980) Endogenous opioids and social behavior. Neurosci Biobehav Rev 4:473–487

Scherer K (1985) Vocal affect signalling: a comparative approach. In: Rosenblatt J, Beer C, Busnel MC (eds) Advances in the study of behavior, vol 15. Academic Press, London New York, pp 189–244

Scherer U, Helfrich H, Scherer KR (1980) Internal push or external pull? Determinants of paralinguistic behavior. In: Giles H, Robinson P, Smith P (eds) Language: social psychological perspectives. Pergamon, Oxford New York, pp 279–282

Schott D (1975) Quantitative analysis of the vocal repertoire of squirrel monkeys *(Saimiri sciureus)*. Z Tierpsychol 38:225–250

Seyfarth RM, Cheney DL (1984) The acoustic features of vervet monkey grunts. J Acoust Soc Am 75(5):1623–1628

Stevenson MF, Poole TB (1982) Playful interactions in family groups of the common marmoset *(Callithrix j. jacchus)*. Anim Behav 30:886–900

Symmes D (1981) On the use of natural stimuli in neurophysiological studies on audition. Hear Res 4:203–214

Winter P, Ploog D, Latta J (1966) Vocal repertoire of the squirrel monkey *(Saimiri sciureus)*, its analysis and significance. Exp Brain Res 1:359–384

Biological Substrates

Ethopharmacology of Vocal Behavior in Primates

JOHN D. NEWMAN [1]

Attempts to experimentally manipulate behavior by chemicals has had a long history. Many of the earlier findings must be viewed with some skepticism and uncertainty, however, due to the questionable mechanisms by which the relatively nonspecific chemicals being employed exerted their effects, as well as to a lack of attention to developing adequate behavioral testing paradigms. A list of some of the earlier work as it applies to primates was provided by Reynolds (1969). With the discovery of chemical ligands that bind to specific chemical receptor subtypes in the brain and other organs, a finer grained analysis of neurochemical systems and their distribution in the brain was initiated which has produced an overwhelming accumulation of neurochemical and neuroanatomical data. The application of this new knowledge to the study of behavioral mechanisms has failed to keep pace. Behavioral pharmacology has attempted to match the rigor of chemical analysis by using operant conditioning methods to impose a level of experimenter control on the behavioral performance of animal subjects. The fruits of this approach have been considerable (see S.D. Iversen and L.L. Iversen 1975, for a general overview; Dews 1972, for the basic methodologies, and the work of Barrett 1985, for a recent review of this approach in one species, the squirrel monkey). However, experimenter control of behavioral performance has been gained at the sacrifice of the questionable relevance of this approach to understanding the chemistry of behavior in natural settings. A few attempts have been made to address the issue of the chemistry of social behavior. Miczek (1983), in his editorial preface to a book on the pharmacology of social behavior in primates, used the term "ethopharmacology" for such studies. Other investigators have employed a more descriptively accurate term for this area of investigation, "sociopharmacology" (e.g., Bellarosa et al. 1980). Such studies represent the opposite extreme to studies using operant conditioning techniques in terms of the number of uncontrolled variables inherent in the study. One of the more complex examples of this approach is the study by Smith and Byrd (1983) on social groups of stumptailed macaques.

Between these two extremes, there is a "middle ground" which, with a few notable exceptions, has largely been ignored. This is the pharmacological study of species-typical behavior patterns. Such behaviors are, by definition, constrained in their variability to that expressed by the members of a single species or group of related species. It is to this category of behavioral pharmacology that I believe the term "ethopharmacology" most apply applies, and the sense in which I will use that term in this work. Vocaliza-

1 Laboratory of Comparative Ethology, National Institute of Child Health and Human Development, P.O. Box 289, NIH Animal Center Bldg. 112, Poolesville, MD 20837, USA

Todt/Goedeking/Symmes (Eds.)
Primate Vocal Communication

tions, as examples of species-typical behavior, are an appropriate choice for pharmacological investigation. They are usually readily detected (even ultrasounds can be detected with the appropriate hardware), and have the potential of being counted as accurately as bar presses. Furthermore, the ability to measure the detailed acoustic structure of vocalizations provides much additional information beyond the rate of behavioral performance that can be of value in interpreting mechanisms of drug action. Yet, few studies of drug effects on vocal behavior have been published. Some studies have noted changes in vocal behavior incidental to documentation of other drug-related behavioral changes. Suomi et al. (1981) reported that daily treatment of young peer-housed macaques with 10 mg kg^{-1} imipramine hydrochloride led to an immediate increase in exploratory behavior and a decrease in "disturbance vocalizations" and stereotyped behavior to near-zero levels when experimentally separated from group members. Placebo-treated control subjects exhibited a typical panic reaction upon separation. Porsolt et al. (1984) found that following single injections of imipramine (doses used were 3.75, 7.5 and 15 mg kg^{-1}), vocalizations actually increased when hand-reared, peer-housed, 1-year-old macaques were separated from group members. Miczek and Gold (1983) reported that following injections of 0.3 mg kg^{-1} d-amphetamine sulfate, half of their squirrel monkeys produced "distresslike" vocalizations while in their social group. Ninan et al. (1982) found that among the behavioral changes following administration of 2.5 mg kg^{-1} beta-carboline ester (an anxiogenic drug) to chaired rhesus macaques, six of seven subjects increased production of "distress vocalizations." Brudzynski and Eckersdorf (1984) found that carbachol injected into the anterior hypothalamus of cats increased the incidence of growling vocalizations. Levine et al. (1984), in a study of the effects of morphine in reducing pain perception in rats, found that a dose of 3 mg kg^{-1} caused a reduction in pain-induced vocalizations.

Other studies have used vocalization as their only behavioral measure. Scott (1974), as part of a long-term study of emotional behavior in dogs, found that diazepam (in doses to 8 mg kg^{-1}) had no effect on separation-induced vocalizations in puppies, but 8 mg kg^{-1} of imipramine hydrochloride significantly reduced vocalizations in this context, showing a peak effect at about 1 h after administration. Thiessen and Upchurch (1981) found that haloperidol and clonidine increased ultrasonic vocalizations of male gerbils, while apomorphine decreased these sounds. Insel et al. (1986) found that administration of 0.5 mg kg^{-1} diazepam resulted in decreased number, power, and frequency of separation-induced vocalizations by rat pups. In the same study, RO 15-1788, a benzodiazepine receptor antagonist, also decreased rat pup isolation calls.

The most thorough studies of drug effects on the rate of vocalizing have been conducted by J. Panksepp and co-workers. Their work has focused mainly on drugs that alter the rate of vocalization induced by separating an infant mammal (excluding primates) or precocial bird (domestic chick) from its family group (referred to as "distress vocalizations" but equivalent to the "isolation calls" of other authors). The role of opiatergic mechanisms in controlling this behavior has received the major share of attention. Panksepp et al. (1978a) found that morphine sulfate produced a dose-dependent drop in the isolation calls of puppies. Herman and Panksepp (1978) extended this finding to infant guinea pigs. Vilberg et al. (1984) demonstrated that beta-endorphin injected intracranially into the cerebral ventricles reduced isolation calls in chicks more effectively than morphine. Panksepp et al. (1978b) used the same testing procedure but administered the opiates peripherally to show that (D-Ala^2) metenkephalin was more effective than beta-endorphin when administered by this route. Panksepp et al. (1982) were able to reverse the vocal-suppressing effects of morphine in chicks by intraventricular administration of a long-acting opiate antagonist, beta-chlornaltrexamine. Other chemical systems shown by this group of investigators to affect rate of chick isolation calls are cholinergics (Sahley et al. 1981) and adrenergics (Rossi et al. 1983). Panksepp (1986) has recently reviewed the findings of his group, along with other studies of behavioral neurochemistry.

It is against this framework that our own studies have been built. We have used as our experimental animal the squirrel monkey *(Saimiri)*, for which extensive descriptions and quantitative analyses of the vocalizations emitted in captivity under a variety of conditions have been published (reviewed by Newman 1985). Our primary focus for use as a behavioral probe is the isolation peep (IP), the species-typical vocalization produced by an individual when separated from familiar conspecifics. The IP is of sufficient loudness and duration and readily distinguishable from other vocalizations to be scored by ear (the listener monitoring through a headphone or loudspeaker the amplified vocal output of the subject being tested). Tests using the IP as a measure of drug action are conducted with the subject visually and acoustically isolated from other monkeys, hence social interactions with conspecifics during testing are eliminated as an uncontrollable variable. The experimenter is also out of view, monitoring the subject's ongoing behavior by closed-circuit television. Previous work from our laboratory has shown that under baseline (no drug or other intervention) conditions, a given adult squirrel monkey produces IPs that are characterized by a stereotyped, consistent, and individually distinctive acoustic structure which varies little between tests (Symmes et al. 1979). This fixed, individually stable structure becomes apparent at about 1 year of age, younger animals showing more variability in vocal output within a single test, and a progressive change in overall IP structure during maturation (Lieblich et al. 1980). For the behavioral pharmacology studies, individuals chosen as subjects are at least 1 year of age (most being considerably older), hence an additional measure of drug action is any deviation from the IPs individually stable acoustic structure.

Our initial studies were aimed at testing whether the effects of opiates on isolation calling in chicks, puppies, and rodents observed by Panksepp and co-workers could also be demonstrated in squirrel monkeys. Naive squirrel monkeys were visually and acoustically isolated from social companions for 15- or 30-min test periods. During these tests, the IP rate (calls per minute) was tabulated, and vocal samples recorded onto a tape recorder for subsequent sound spectrographic analysis. Tests were made under the following conditions: no injections (subject is coaxed into a wire-mesh cage and transferred immediately to the test room); systemic (parenteral or subcutaneous) injection of morphine sulfate (5 or 10 mg kg^{-1} body weight) 1 min prior to testing; systemic injection of naloxone hydrochloride (0.5 mg kg^{-1}) 15 min after the morphine, and 1min prior to testing; saline vehicle alone injected 1 min prior to testing. For each subject, tests were conducted at weekly intervals as a series: no injection (30-min test); morphine (15-min test) followed immediately by naloxone (15 min); saline (30-min test). In some weekly series, saline was substituted for naloxone, while in other series, injection of 0.5 mg kg^{-1} naloxone was administered in absence of the prior morphine injection and the subject tested 1 min later. Our results, which have been previously published only in an abstract (Newman et al. 1982), can be summarized as follows. We found that systemic administration of morphine sulfate in doses of 5 or 10 mg kg^{-1} significantly reduced isolation call production in adult squirrel monkeys. Typically, a subject remained totally silent throughout the test. Examination of subjects following the morphine test showed them to be alert and responsive to external stimulation. Subjects were then given either naloxone or saline and tested for another 15 min. Naloxone-injected animals began vocalizing within 4 min, and continued for variable lengths of time, whereas animals injected with saline were silent throughout this second test period. A typical experimental series is shown in Fig. 1. The subject (934) called steadily during a 30-min control test, gradually tapering off at the end of this time. Under morphine, this subject was totally silent. Subsequent administration of naloxone caused a resumption of calling, in this particular test achieving a higher rate than during the control experiment. The potentiation of the IP rate over control levels by naloxone was not a consistent or repeatable finding in these experiments, but did occur in a more recent study (Harris and Newman 1986). A summary of the results for all three subjects is shown in Fig. 2. As indicated, treatment with morphine resulted in a significant drop in isolation calling, an effect that was reversed by the subsequent naloxone injection. These experiments provide the first evidence that opiatergic mechanisms play a role in mediating isolation call production in a primate, as had been demonstrated in rodents and chicks by Panksepp and co-workers.

The possibility that the catecholamines might affect IP production in squirrel monkeys came from prior work demonstrating that clonidine (an alpha-2 adrenergic receptor agonist) caused a reduction in separation-induced distress vocalizations in chicks, an effect that was reversed by administering the alpha-2 antagonist yohimbine (Rossi et al. 1983). There also was the proposal that imipramine, a drug important in the

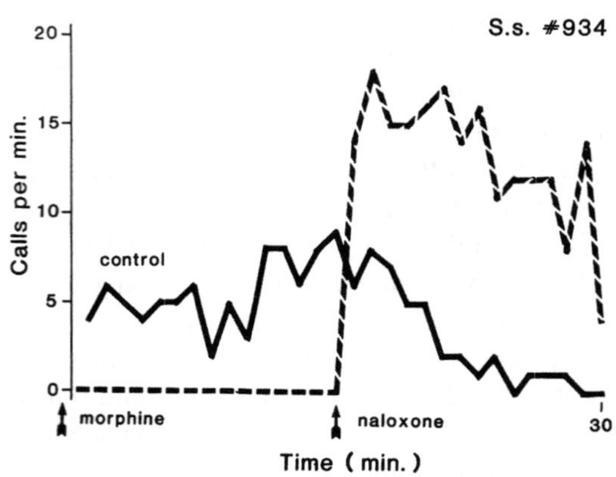

Fig. 1. Typical experiment testing exogenous opiate modulation of squirrel monkey isolation calls. Subject *934* received 10 mg kg^{-1} morphine sulfate i.m., and failed to utter any isolation calls over a 15-min period of separation from conspecifics. The subject was then recaptured, given an i.m. injection of 0.5 mg kg^{-1} naloxone hydrochloride, then immediately returned to the isolation testing room. The animal began emitting isolation calls within a few minutes, and continued to call at a high rate for the remainder of the 15-min test. A typical control experiment (conducted 1 week from drug administration), in which the subject received only a saline injection, is shown superimposed on the drug test

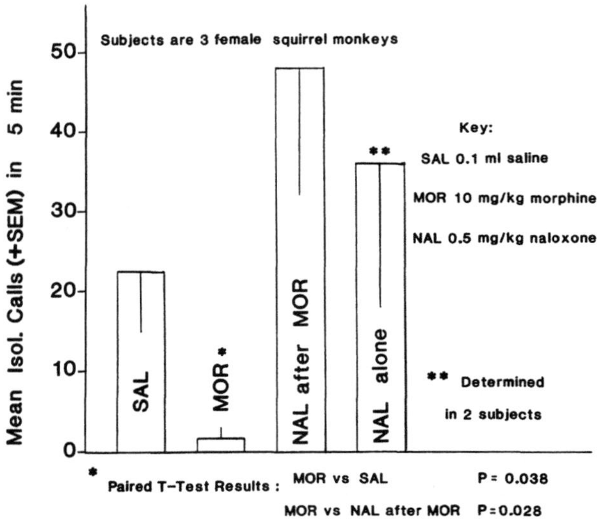

Fig. 2. Summary of results of tests for opiate modulation of isolation calls for three adult squirrel monkeys. Calling rate values shown are for first 5 min of each test

clinical management of separation anxiety, acted by suppressing the activity of catecholamine-containing neurons, particularly those in the locus coeruleus (Nyback et al. 1975). This nucleus is a prominent source of nerve fibers contributing to the noradrenergic system ascending to the forebrain. Scott (1974) had reported that imipramine over a dose range of 2–8 mg kg^{-1} administered to 8-week-old beagle puppies produced a graded depression of isolating calling. Maximum suppression of calling rate occurred 1 to 2 h after drug administration. By 4 h, the drug effect could no longer be demonstrated. Harris and I recently (Harris and Newman 1987) tested imipramine's effect on isolation call production in squirrel monkeys. In four adults, we found that 10 mg kg^{-1} imipramine reliably eliminated or significantly reduced isolation call production. As Scott had reported in his beagle study, we found that suppression of isolation calls occurred when the interval between drug administration and testing was 1 h. If imipramine was acting through its interplay with the adrenergic system, it should be possible to manipulate isolation call production by manipulating the activity of this system. Our approach to this was to test if an alpha-2 receptor agonist (clonidine) could reduce isolation calling, and, if so, whether a specific alpha-2 antagonist (yohimbine) could block or reverse the suppression of vocal behavior. That is, in fact, what we have found, and reported in 1985. Subsequently, we have found that yohimbine administered alone causes a dose-dependent *increase* in isolation calling (Harris and Newman 1987). This ability to manipulate isolation call rate up or down by administering drugs that specifically reduce or enhance activity at the alpha-2 adrenergic receptor strongly implicates the adrenergic component of the catecholamine system in the regulation of this behavior.

Although the underlying emotional states accompanying increases in IP calling rate are presently unknown, the fact that calling rate correlates positively with activity (Fig. 3) suggests that as the production of isolation calls increases, the level of arousal also increases. Given this possibility, it follows that any drug that leads to increased isolation calling might also lead to increased production of other vocalizations associated

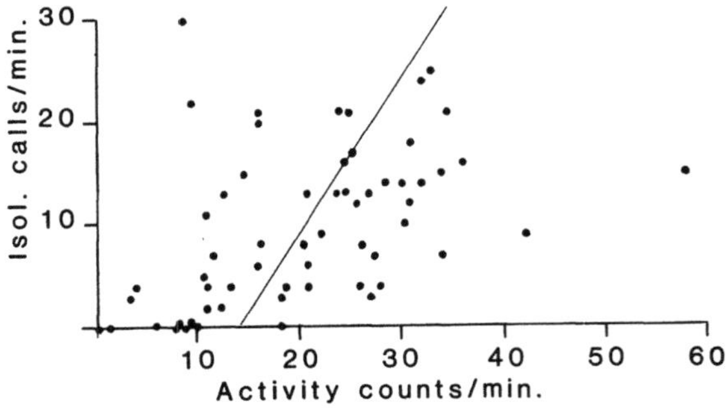

Fig. 3. Scatter plot of regression analysis comparing isolation calling rate vs activity. Data are taken from four male squirrel monkeys, each separated from conspecifics for 15 min. Activity was scored as one count for each quadrant within test cage crossed. Slope of regression line = 0.6605; intercept = 14.049. Correlation coefficient = 0.4586 (significant at a p value of < 0.01)

with increased arousal level. The best test of this would be situations in which an alarming stimulus is present, where the typical reaction involves both the production of alarm calls and other signs of increased arousal (e.g., piloerection, sudden attempts to escape). I have developed a standardized testing procedure that permits quantification of the strength and persistence of alarm reactions. The procedure involves repetitive presentations of the "alarm stimulus" (a hand puppet covered with artificial fur and with loosely hanging arms and legs) to a monkey alone in the test cage. Each presentation lasts 30 s, with an intervening 30 s when the puppet is withdrawn from the subject's view. Each test consists of five or ten stimulus presentations. Prior to testing, the subject is given an intramuscular injection either of a drug or sterile saline (as a control). In control tests, subjects typically react to this stimulus by leaping to a far corner of the test cage and emitting repeated calls with a sharp onset and rich harmonics variously termed *yaps, barks*, and *cackles* (Winter et al. 1966; Winter 1968; Schott 1975; Newman 1985). This reaction quickly habituates, however, so that by the fourth or fifth trial, the subject may not vocalize at all. Glowa and I (Glowa and Newman 1986) have discovered that benactyzine hydrochloride, an anticholinergic drug, greatly potentiates this alarm reaction, both in increasing the number of calls per trial, and in extending the number of trials over which the subject shows an alarm reaction. Benactyzine hydrochloride is known to enter the brain following peripheral administration and block activity at cholinergic receptors of the muscarinic type. We found that benactyzine administered over a dose range of 0.03–3.0 mg kg^{-1} caused a dose-dependent increase in yaps, barks, and cackles. In most cases, call rates decreased to a baseline level after the stimulus was withdrawn.

As a follow-up to these experiments, I tested the relative effectiveness of benactyzine and yohimbine on alarm calling and isolation calling in a new group of squirrel monkeys. It is quite clear, as indicated in Fig. 4, that the two drugs are selective as to which class of vocalization they potentiate, since no isolation calls are produced following benactyzine and no alarm calls are produced following yohimbine.

1 Discussion and Conclusions

Data from several pharmacological studies reviewed here indicate that calling rate of certain species-typical vocalizations of squirrel monkeys can be manipulated either up-

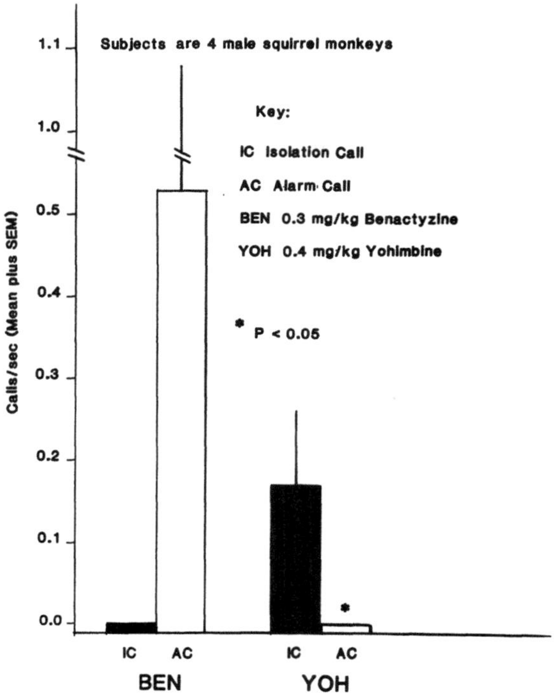

Fig. 4. Comparison of effects of a cholinergic (benactyzine) and adrenergic (yohimbine) antagonist on calling rate of adult squirrel monkeys. Each subject was administered one or the other drug and tested in one of two standard tests (separation from conspecifics or reaction to an alarm stimulus; details described in text). Tests were run at least 1 week apart. The figure shows that benactyzine selectively enhances alarm calls but not isolation calls, whereas yohimbine has the opposite effect, selectively enhancing production of isolation calls but not alarm calls

ward or downward by peripheral administration of the appropriate drug. In the case of vocal alarm, there is a clear relationship between dosage of the effective drug (benactyzine) and response strength, suggesting the involvement of a central cholinergic mechanism that is linked to the production of alarm calls in a relatively direct and uncomplicated way. With respect to isolation calls, however, the chemical control mechanisms appear to be more complicated, as suggested by a nonlinear dose-response curve for yohimbine, and the fact that both alpha-adrenergic and opiatergic drugs can mediate isolation call rate. We have recently found that yohimbine and naloxone act synergistically to increase production of another vocalization, the twitter, over calling rates to either drug alone (Harris and Newman 1986). This synergism suggests that a more complex chemical mechanism modulates production of calls used in social contact (isolation calls and twitters) than may be the case for the chemical mediation of alarm calls (yaps, barks, and cackles). It would not be surprising if future research demonstrates that this is the case, since alarm reactions have an almost reflexlike quality to them (but usually habituate quickly), whereas vocal interactions mediating group contact require more complex factors to initiate them, such as recognition of other troop members and distance of the vocalizer from the troop.

2 Contribution of Ethology and Bioacoustics to Behavioral Pharmacology

At the outset of this chapter, an argument was made for the use of species-specific vocalizations as probes for studying pharmacological effects on behavior. Such a change in the orientation of behavioral pharmacologists would no doubt require their close co-operation with the zoologists and comparative psychologists most familiar with the vocal repertoires of primates and other mammals suitable for laboratory experimentation. Zoologists would benefit from this interaction by gaining new insights into the physiological mechanisms controlling natural behavior, as well as by learning how to promote the occurrence of vocal behavior that might otherwise be only rarely observed in a laboratory setting. One major role of the bioacoustics expert in these interdisciplinary behavioral pharmacology experiments would be to perform the detailed structural analyses of drug-induced vocalizations needed to confirm whether they show systematic differences from similar vocalizations occurring without drug intervention. Demonstrating the normalcy of drug-induced vocal behavior is, of course, important in interpreting the mechanism and possibly also the central site of action of the pharmacological agents being used.

References

American Psychiatric Association (1980) Diagnostic and statistical manual of mental disorders, 3rd ed. Washington DC

Barrett JE (1985) Behavioral pharmacology of the squirrel monkey. In: Rosenblum LA, Coe CL (eds) Handbook of squirrel monkey research. Plenum, New York London, pp 315–348

Bellarosa A, Bedford JA, Wilson MC (1980) Sociopharmacology of d-amphetamine in *Macaca arctoides*. Pharmacol Biochem Behav 13:221–228

Brudzynski SM, Eckersdorf B (1984) Inhibition of locomotor activity during cholinergically-induced emotional-aversive response in the cat. Behav Brain Res 14:247–253

Dews PB (1972) Assessing the effects of drugs. In: Myers RD (ed) Methods in psychobiology, vol 2. Academic Press, London New York, pp 83–124

Glowa JR, Newman JD (1986) Benactyzine increases alarm call rates in the squirrel monkey. Psychopharmacology 90:457–460

Harris JC, Newman JD (1985) Alpha-2 adrenergic receptor involvement in adult male squirrel monkey vocal behavior. Soc Neurosci Abstr 11 (1):5

Harris JC, Newman JD (1986) Synergistic effects of alpha-2 adrenergic and opiate receptor blockade on vocalization in squirrel monkeys. Soc Neurosci Abstr 12 (2):1133

Harris JC, Newman JD (1987) Mediation of separation distress by alpha-2 adrenergic mechanisms in a nonhuman primate. Brain Res 410:353–356

Herman BH, Panksepp J (1978) Effects of morphine and naloxone on separation distress and approach attachment: evidence for opiate mediation of social affect. Pharmacol Biochem Behav 9:213–220

Herman BH, Panksepp J (1981) Ascending endorphin inhibition of distress vocalization. Science 211:1060–1062

Insel TR, Hill JL, Mayor RB (1986) Rat pup ultrasonic isolation calls: possible mediation by benzodiazepine receptor complex. Pharmacol Biochem Behav 24 (5):1263–1267

Iversen SD, Iversen LL (1975) Behavioral pharmacology. Oxford Univ Press, Oxford New York

Levine JD, Feldmesser M, Tecott L, Gordon NC, Izdebski K (1984) Pain-induced vocalization in the rat and its modification by pharmacological agents. Brain Res 296:121–127

Lieblich AK, Symmes D, Newman JD, Shapiro M (1980) Development of the isolation peep in laboratory-bred squirrel monkeys. Anim Behav 28:1–9

McKinney WT Jr, Eising RG, Moran EC, Suomi SJ, Harlow HF (1971) Effects of reserpine on the social behavior of rhesus monkeys. Dis Nerv Syst 32:735–741

Miczek KA (ed) (1983) Ethopharmacology: primate models of neuropsychiatric disorders. Liss, New York

Miczek KA, Gold LH (1983) d-Amphetamine in squirrel monkeys of different social status: effects on social and agonistic behavior, locomotion, and stereotypies. Psychopharmacology 81:183–190

Newman JD (1985) Squirrel monkey communication. In: Rosenblum LA, Coe CL (eds) Handbook of squirrel monkey research. Plenum, New York London, pp 99–126

Newman JD, Murphy MR, Harbaugh CR (1982) Naloxone-reversible suppression of isolation call production after morphine injections in squirrel monkeys. Soc Neurosci Abstr 8:940

Ninan PT, Insel TM, Cohen RM, Cook JM, Skolnick P, Paul SM (1982) Benzodiazepine receptor-mediated experimental "anxiety" in primates. Science 218:1332–1334

Nyback HV, Walters JR, Aghajanian GK, Roth RH (1975) Tricyclic antidepressants: effects on the firing rate of brain noradrenergic neurons. Eur J Pharmacol 32:302–312

Panksepp J (1986) The neurochemistry of behavior. Annu Rev Psychol 37:77–107

Panksepp J, Herman B, Conner R, Bishop P, Scott JP (1978a) The biology of social attachments: opiates alleviate separation distress. Biol Psychiat 13(5):607–618

Panksepp J, Vilberg T, Bean NJ, Coy D, Kastin AJ (1978b) Reduction of distress vocalization in chicks by opiate-like peptides. Brain Res Bull 3:663–667

Panksepp J, Siviy S, Normansell L, White K, Bishop P (1982) Effects of beta-chlornaltrexamine on separation distress in chicks. Life Sci 31:2387–2390

Porsolt RD, Roux S, Jalfre M (1984) Effects of imipramine on separation-induced vocalizations in young rhesus monkeys. Pharmacol Biochem Behav 20:979–981

Raleigh MJ, Brammer GL, McGuire MT, Yuwiler A (1985) Dominant social status facilitates the behavioral effects of serotonergic agonists. Brain Res 348:274–282

Redmond DE, Maas JW, Kling A, DeKirmengian H (1971) Changes in primate social behavior after treatment with alpha-methyl-para-tyrosine. Psychosom Med 33:97–113

Reynolds HH (1969) Nonhuman primates in the study of toxicological effects on the central nervous system: a review. Ann NY Acad Sci 162:617–629

Rossi J III, Sahley TL, Panksepp J (1983) The role of brain norepinephrine in clonidine suppression of isolation-induced distress in the domestic chick. Psychopharmacology 79:338–342

Sahley TL, Panksepp J, Zolovick AJ (1981) Cholinergic modulation of separation distress in the domestic chick. Eur J Pharmacol 72:261–264

Schott D (1975) Quantitative analysis of the vocal repertoire of squirrel monkeys *(Saimiri sciureus)*. Z Tierpsychol 38:225–250

Scott JP (1974) Effects of psychotropic drugs in separation distress in dogs. In: Neuropsychopharmacology. Proc IX Congr CINP, Paris. Excerpta Medica, Amsterdam, pp 735–745

Smith EO, Byrd LD (1983) Studying the behavioral effects of drugs in group-living nonhuman primates. In: Miczek KL (ed) Ethopharmacology: primate models of neuropsychiatric disorders. Liss, New York, pp 1–31

Suomi SJ, Seaman SF, Lewis JK, DeLizio RD, McKinney WT Jr (1978) Effects of imipramine treatment on separation-induced social disorders in rhesus monkeys. Arch Gen Psychiat 35:321–325

Suomi SJ, Kraemer GW, Baysinger CM, DeLizio RD (1981) Inherited and experiential factors associated with individual differences in anxious behavior displayed by rhesus monkeys. In: Klein DF, Rabkin J (eds) Anxiety research and changing concepts. Raven, New York, pp 179–199

Symmes D, Newman JD, Talmage-Riggs G, Lieblich AK (1979) Individuality and stability of isolation peeps in squirrel monkeys. Anim Behav 27:1142–1152

Thiessen DD, Upchurch M (1981) Haloperidol and clonidine increase, and apomorphine decreases ultrasonic vocalizations by gerbils. Psychopharmacology 75:287–290

Vilberg TR, Panksepp J, Kastin AJ, Coy DH (1984) The pharmacology of endorphin modulation of a chick distress vocalization. Peptides 5:823–827

Winter P (1968) Social communication in the squirrel monkey. In: Rosenblum LA, Cooper RW (eds) The squirrel monkey. Academic Press, London New York, pp 235–253

Winter P, Ploog D, Latta J (1966) Vocal repertoire of the squirrel monkey *(Saimiri sciureus)*, its analysis and significance. Exp Brain Res 1:359–384

Neural Basis of Signal Detection

Peter Müller-Preuss [1]

When reviewing pertinent literature, it becomes evident that the auditory pathway is quite well known at its structural level. In comparison with other sensory systems, it has some unique features: an extraordinary divergence in the number of cells, for instance, from a relatively few peripheral ones up to millions of neurons on the cortical level. This is probably due to the fact that the auditory system has to deal more with temporal than spatial aspects (in comparison with the visual system). Another unique feature of the system is that, at each level, the particular nuclei are divided into many subgroups, and it is also worth noting that numerous connections exist between these nuclei, even across the midline. These structural properties suggest a great complexity of connections within the pathway and consequently, studies of its functional properties are burdened with many problems. Regarding the problem of complex relationships, the greatest number of investigations were concentrated on the processing mechanisms of simple acoustic stimuli, such as tones, noise, clicks, AM and FM sounds, through the auditory pathway. By using such sounds, which are clearly described on the basis of its spectral, temporal, and spatial parameters, the investigator has better control of experimental conditions and will therefore obtain more reliable results enabling him to arrive at more convincing interpretations. Studies based on such conditions contributed much to our understanding of the basic mechanism through which the auditory system receives and encodes particular physical parameters of acoustic stimuli (Aitkin et al. 1984).

Psychometric studies have described the auditory channel in investigating the threshold-frequency relationships as follows: primates receive acoustic stimuli over a very broad range of frequencies, from 50 Hz to 60 kHz (Stebbins 1980). The size of the head primarily accounts for species-specific differences: the smaller the head, the higher the detectable frequency, and vice versa. Environmental factors, e.g. forestal or non-forestal habitat, play only a minor role, influencing more the band of frequency used for a particular vocal behavior. Most vocal signals of primates are centered around the species-specific area of highest sensibility and rarely affect the borderlines of the threshold curves. So there is an overall correspondence between audiogram and vocal repertoire. On the neural level, characteristic of most neurons in the auditory pathway is that they respond, to varying degrees, selectively to the frequency of an (simple) acoustic stimulus. The measure for this selectivity is the so-called tuning curve of a neuron which defines its threshold as a function of the stimulus frequency. Theoretically,

1 Max-Planck-Institut für Psychiatrie, Kraepelinstr. 2, 8000 München 40, FRG

Todt/Goedeking/Symmes (Eds.)
Primate Vocal Communication

integration of the tuning curves of all neurons should match the psychometrically evaluated threshold/frequency relationships.

Another feature of the auditory nuclei is that their neurons are spatially arranged according to their frequency response characteristics. The so-called tonotopic organization clearly reflects the structural organization of the peripheral sense organ, the cochlea. This tonotopic organization is found at all levels of the auditory pathway of primates, not only in the primary areas, but also in regions which are assumed to be involved in more secondary processing mechanisms. For example, in the first station of the pathway, the cochlear nucleus, tonotopic organization has been shown to exist in all three subnuclei, i.e., the dorsal, the antero- and posteroventral part. Higher up in the system, the arrangement occurs in the three subunits of the superior olivary complex, namely, the nuclei of the lateral lemniscus, the central nucleus of the inferior colliculus, the lateral part of the medial geniculate body, and in particular regions of the auditory cortex. The fact that the tonotopic organization is frequently repeated within the system indicates that such a basic device is used for a variety of analytic procedures established for the reception and detection of the numerous "features" complex signals consist of. Other properties of the system possibly influencing the detectability of fine differences between the various types of a vocal repertoire are the abilities to distinguish not only the degrees in the changes of frequency but also in its intensity. It has been shown that there are species-specific differences regarding such abilities. Humans, for example, can resolve frequency-difference limens down to 4 Hz, whereas macaques can resolve them only to 16 Hz. With regard to intensity difference limens, monkeys perform as good as humans if the intensity increases, but become worse if the intensity decreases (1 dB vs 4.4 dB) (Sinnot et al. 1985). Such limiting factors may play a role in the detection of vocal signals and influence the ability to receive one type better than another.

Again at the neuronal level, there are studies which report on an increasing complexity of neural responses evoked by stimuli changing in intensity in correspondence to the ascent of the auditory pathway. Within the anterior ventral cochlear nucleus the responses are similar to those of the fibers of the eighth nerve: an increasing intensity evokes an increasing discharge rate up to the level of saturation. Such relatively simple monotonic rate/level functions are found along the whole pathway. Nevertheless, already in the dorsal part of the cochlear nucleus nonmonotonic rate/level functions have been found: a particular sound intensity, not necessarily the strongest, evokes the strongest response. Cells displaying such nonmonotonic functions are considered to detect and encode specific levels of intensity. Ascending the pathway, they increase in number and are even found in the more "secondary regions".

Neurons whose activity can reflect specific sound pressure levels can also serve to detect the location of a signal in space. Depending on the location of the sound source, the size of the head and, consequently, the shadow caused by it for one ear, sounds having high frequencies will arrive at different intensities at the ears. Via a binaural decoding mechanism, such interaural intensity differences (IID) serve for the detection of a sound's location. Binaural cells reacting differently (for example, excitatory-excitatory or excitatory-inhibitory or excitatory-no response) are found first at the level of the superior olivary complex. There the lateral nucleus of this complex uses more intensity differences (= high frequencies), whereas the medial nucleus responds more to the different times of arrival of a stimulus. Such a mechanism using interaural time differences (ITD) requires neurons which respond to low frequencies. Again, neurons displaying binaural response mechanisms are found along all stations of the pathway up to the cortex. Embedded within the particular frequency response areas, collicular neurons, for example, which use IID, lie more ventromedially, neurons which use ITD lie more dorsolaterally within the central nucleus of this station of the auditory midbrain. For the cortex, another topological arrangement has been demonstrated in cats: different types of binaural responding cells (EE, EI) are organized in stripes perpendicular to the isofrequency lines (Middlebrooks et al. 1980).

All these results, which of course are incomplete (e.g., cells also responding selectively to AM or FM stimuli have been recorded), nevertheless contributed enormously to the picture we now have of the neuronal mechanisms of the reception and detection of acoustic stimuli with simple physical structures. A good example of how useful this knowledge can be for medical application is the fact that it was used for the development of cochlear implants which are a great help to humans suffering from severe malfunction of their auditory sense. How much do these data support our efforts to understand the neuronal bases of those signals which play a role in vocal communication of primates? It is obvious that most of the vocal signals of primates do not have a simple physical structure. Their vocal repertoire contains broadband signals which are more or less also modulated in frequency and amplitude. In the last 2 or 3 decades, auditory physiologists concentrated more and more on solving the problems connected with the neuronal processing mechanisms of species-specific vocalizations or of the so-called biologically relevant sounds within the auditory pathway. Neurophysiological recording experiments using such sounds as test stimuli have been carried out on various species, including insects, frogs, birds, rats, cats and, despite their obviously complex acoustic behavior, primates. With regard to the complexity of behavior, it is easier to obtain results from lower animals (e.g., insects, frogs) than from mammals, especially primates. This reflects the fact that the repertoire of sounds used by lower animals is simpler and the relationships between sounds and behavior less variable. In this regard, reference is made to the clear relationship between neuronal responses and the phonotactic behavior in crickets or the reactions of those neurons in the auditory pathway of frogs which encode mating and release calls in this species. Here the responses to natural calls can be predicted by considering responses to simple artificial stimuli; in other words, the reaction to vocalizations fits into the general response characteristic of the neurons.

Predictability of responses to species-specific vocalizations has also been reported for the auditory pathway of primates, but there is an increasing number of pathway-ascending cells whose call-evoked activity does not fit into the general response properties. This is not very surprising in view of the high complexity of the system and of the vocalizations which have to be identified. In contrast to the lower animals mentioned above and also to birds, only very few units responding selectively to a particular vocalization have been found in the auditory pathway of a primate (Newman and Wollberg 1973). These findings imply that the concept of a so-called pontifical or grandmother cell where all call-evoked excitatory or inhibitory neural activities converge gets only weak experimental support.

Other approaches were used where specific sets of call parameters (often defined as features), not the whole call, were considered as stimuli. But in no recording study carried out on primates has it been possible to demonstrate that the basis of processing relies on definite call parameters insofar as these parameters selectively evoke the strongest and most stable reaction. So we have a rather reliable picture concerning the neural bases of reception and detection of simple artificial stimuli, but we are far from understanding how, at the neural level, the auditory system processes complex stimuli, such as species-specific vocalizations. Of course, there exist already a large collection of data regarding the response pattern of auditory neurons in primates evoked through species-specific calls. But mostly the knowledge obtained is very general. We learned that they deliver information they can respond to, that they respond differently to different stimuli, that responses vary in time, and that the behavioral state also influences the response. The following considerations may explain the difficulties encountered in evaluating satisfactory results:

1. The stimulus repertoire selected is not appropriate;
2. The recording situation differs too much from the natural context;
3. The bases of signal detection cannot be evaluated on the single-unit level.

With regard to the last two points, new approaches are thinkable, but difficult technical problems would be involved. In a group of monkeys, for instance, the activity of neurons could be recorded telemetrically, which should be sufficient for an investigation of the mechanisms of acoustic communication. But it would not solve the problems in connection with single-unit recordings. There are several studies in progress, where the activities of up to 20 neurons are recorded simultaneously by using a multichannel recording system. But aside from the fact that these studies are at their very beginning, one can well imagine the problems which arise if these activities are not only recorded simultaneously but also transmitted telemetrically, amplified, and stored. Obviously, these problems will not be overcome in the near future, so that neurophysiologists can contribute only the limited interpretations of single-unit data obtained from isolated animals.

Now, regarding the first point, appropriateness of the stimulus repertoire: The vocalizations were selected on the basis of certain criteria, e.g. whether they were representative for the species, contained particular physical parameters and, if possible, were of biological significance. The selection was made with the aid of sound spectrograms and oscillograms. As already mentioned, most vocalizations are of a complex nature, with many parameters possibly causing the neural reaction. It will often be difficult, therefore, to decide which part of a call triggers the neuron. To learn more about the mechanisms of signal detection at the neuronal level, a tremendous amount of calls or call parts has to be tested on a tremendous amount of neurons along the pathway.

We think that there is one approach not requiring such a great number of stimuli: if we could determine which part (in physical terms) of a call is used by the animals in their communicative behavior. So far, when looking for the neural bases of the detection of signals, we could concentrate only on those stimuli which are of communicative relevance. However, recently a concept based on behavioral methods and meanwhile supported by experimental data, has been developed by which essential functions or cues within a vocal repertoire can be detected (Maurus et al. 1984). Our new approach is to use only those calls or call parts as stimuli in neurophysiological recording experiments which fulfill the above mentioned conditions. More specifically, cackle calls showing no frequency modulation, selected according to definite amplitude changes, were correlated with the activity of neurons of the auditory midbrain (colliculus inferior) and thalamus (medial geniculate body).

Before presenting several examples, some methodological remarks are necessary: experiments were carried out on conscious squirrel monkeys restrained in a contour padded chair with fixed heads. Single-unit recordings were done with the aid of a microdrive and tungsten microelectrodes. Computer-generated vocalizations were presented to the monkeys in a free-field situation. The vocalizations were generated from digitized tokens obtained from those recorded telemetrically during the behavioral experiments. To register the influence of intensity in general, the sound pressure level was varied in steps of 10 dB. Highest intensity (= 0 dB attenuation) was 65–75 dB. Furthermore, only units with broad tuning curves were considered for analysis. For further details of methods, see also Müller-Preuss and Maurus (1985).

Many neurons of our sample displayed simple response properties where only the level of the particular amplitudes seems to be encoded. An example of those neurons is given on the left side (A) of Fig. 1. However, a relatively great number of neurons already in the midbrain and also from the thalamus showed response patterns which support evidence that parts of the auditory pathway are able to process specific changes in the course of the intensity of a call. On the right side (B) of Fig. 1, a unit is shown where, despite the same level of intensity, only the first and the third amplitude change cause a response, the second does not. A further example, shown in Fig. 2, demonstrates a cell with a very broad, dynamic range (> 50 dB) which seems to act as a level detector at the beginning of a call, independently of how the call starts: with a high or a small amplitude. But in this sample it can already be seen that the kind of the beginning amplitude influences the following amplitude changes: on the left side, the third change evokes a response, on the right side it does not. The third example (Fig. 3) shows the responses to call type A and A′ where the position of the first amplitude change defines the reaction. In the inversed call (A′), where this change is now at the third position, no reaction can be seen. A further neuron (Fig. 4) demonstrates that the direction and/or position has an influence on the activity. In B the cell responds after the end of the

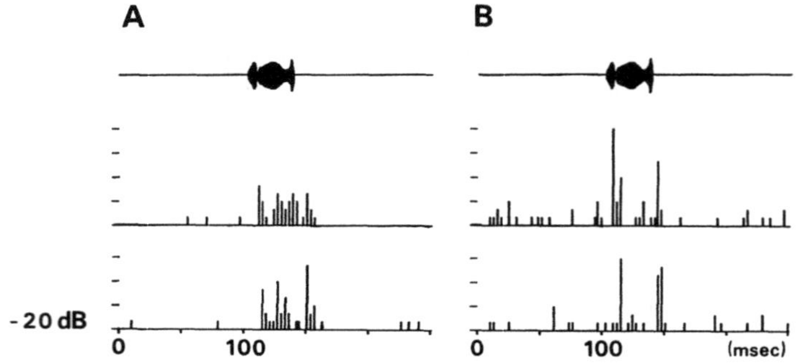

Fig. 1. Response patterns of two collicular cells (**A, B**) as shown by peri-stimulus time histograms. Above the particular PSTHs the course of amplitude changes is schematically outlined. Binwidth 6 ms

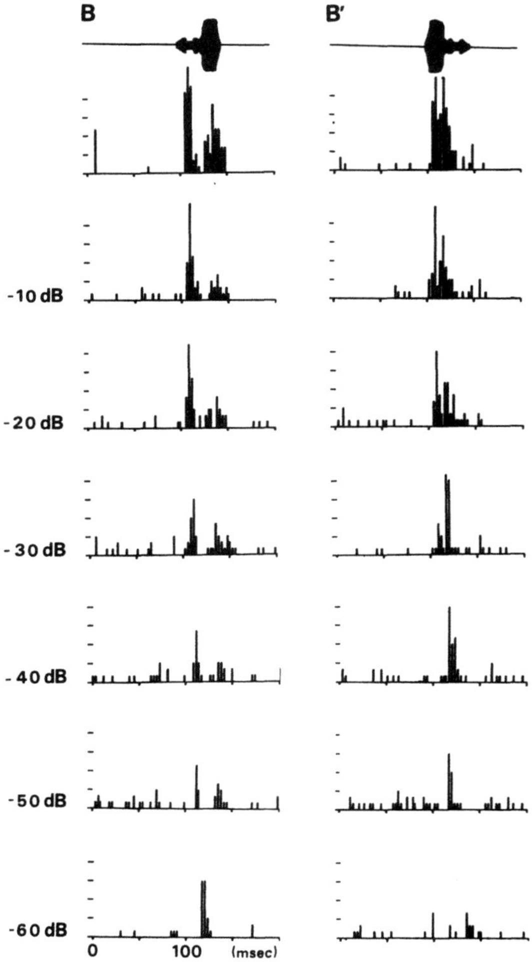

Fig. 2. Response pattern of a collicular cell, evoked through a call first presented with the original amplitude course (**B**), and then presented with the inversed amplitude (**B'**). Binwidth 3 ms

call, a fact which may be interpreted as meaning that the decrement of intensity is responsible for the reaction. In B', where the same amplitude change is now at the beginning of the call with an increment of intensity, the comparable reaction is now very weak.

A last example (Fig. 5), from a collicular cell, substantiates the demonstrated observations: The reactions to three calls (A, B, C) presented at three intensities are shown. This neuron displays differing responses to the particular vocalizations. Relatively high amplitude changes evoked a response, but not lower amplitude changes or higher amplitude levels. The first change evoked a greater reaction than the second or third one (see example A or B). This happened also when the amplitude level of a subsequent change was high (B). On the other hand, when there was no immediately preceding amplitude change, a call beginning with a high amplitude caused a great discharge rate (C). With decreasing reaction to the first change, succeeding changes evoked responses of an increasing discharge rate.

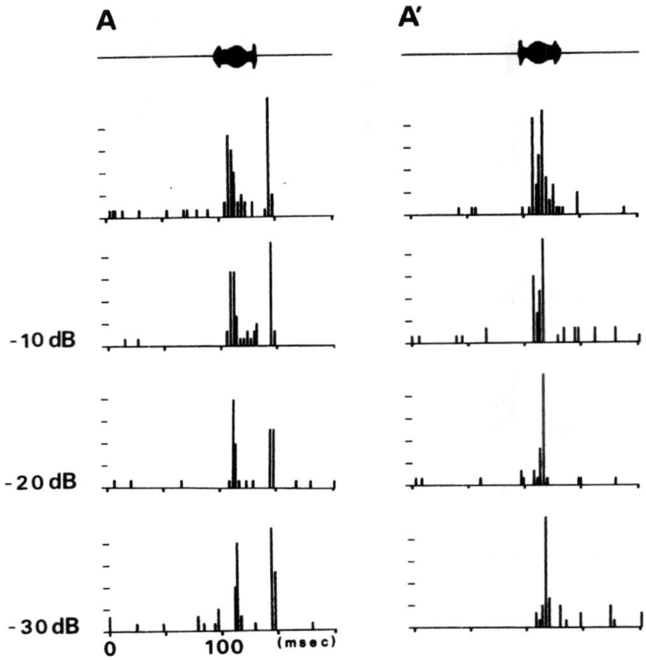

Fig. 3. Response pattern of a collicular cell, evoked through the original (**A**) and inversed (**A′**) call. Binwidth 3 ms. This example demonstrates the influence of the position of a particular amplitude change within the call

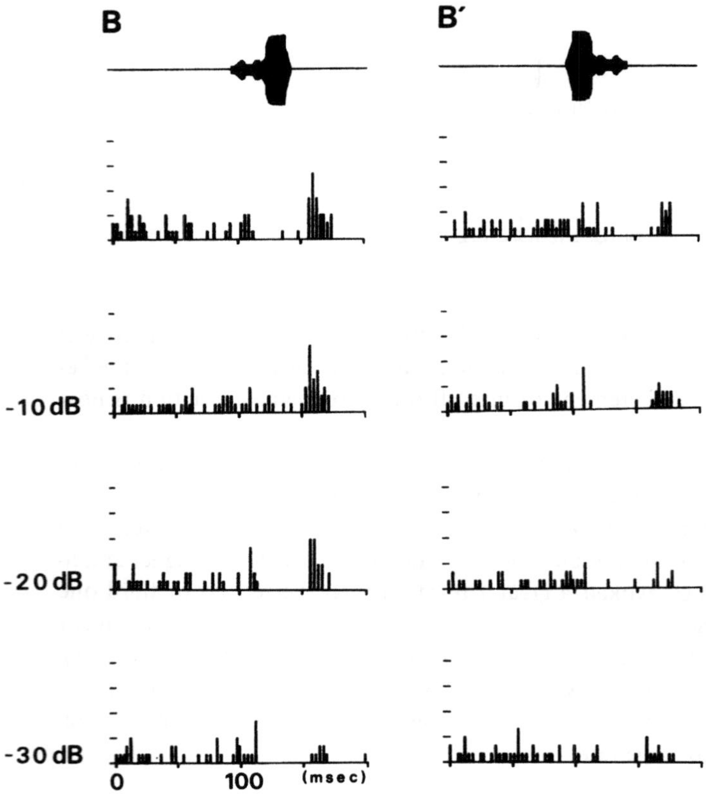

Fig. 4. Response pattern of a collicular cell, evoked by the original (**B**) and inversed (**B′**) call. Binwidth 3 ms. The example demonstrates the influence of the direction of an amplitude change

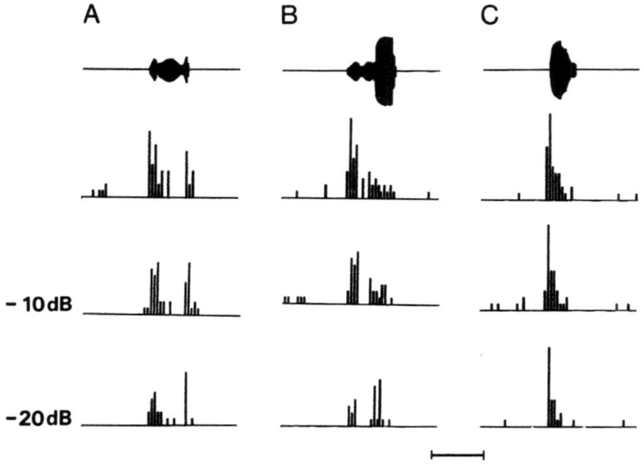

Fig. 5. Responses of a collicular cell to three calls (**A, B, C**) with different amplitude courses. Binwidth 3 ms; *bar* = 50 ms. Latency of first evoked activity 7 ms. In this particular figure latency has been compensated for better comparison of call components with neural activity

The examples shown indicate that specific amplitude changes can be processed independently of amplitude levels and that the position and, furthermore, the direction (increment vs decrement) of a change within a call can be encoded. The results give evidence that call components indicative of being relevant to intraspecific communication elicit the neuronal activity necessary for perception.

The "new approach", outlined above, as supported by first results also described, seems to be an additional tool in evaluating the neuronal bases of signal detection. It is an approach which relies on both neurophysiological as well as behavioral methods.

Acknowledgment. Supported by the Deutsche Forschungsgemeinschaft SFB 204/11.

References

Aitkin LM, Irvine DRF, Webster WR (1984) Central neural mechanisms of hearing. In: Smith D (ed) Handbook of physiology. Bethesda, pp 675–737

Maurus M, Streit KM, Geissler B, Barclay D, Wiesner E, Kühlmorgen B (1984) Categorical differentiation in amplitude changes of squirrel monkey calls. Lang Commun 5:195–208

Middlebrooks JC, Dykes RW, Merzenich MM (1980) Binaural response-specific bands in primary auditory cortex (A1) of the cat: topographical organization orthogonal to isofrequency contours. Brain Res 181:31–48

Müller-Preuss P, Maurus M (1985) Coding of call components essential for intraspecific communication through auditory neurons in the squirrel monkey. Naturwissenschaften 72:437

Newman JD, Wollberg Z (1973) Multiple coding of species-specific vocalizations in the auditory cortex of squirrel monkeys. Brain Res 54:287–304

Sinnot JM, Petersen MR, Hopp SL (1985) Frequency and intensity discrimination in humans and monkeys. J Acoust Soc Am 78:1977–1985

Stebbins WC (1980) The evolution of hearing in mammals. In: Popper AN, Fay RR (eds) Comparative studies of hearing in vertebrates. Springer, Berlin Heidelberg New York, pp 421–436

Central Control of Monkey Calls

UWE JÜRGENS [1]

The central nervous control of monkey calls differs in several respects from that of human speech. This is mainly due to three reasons. Firstly, monkey calls are genetically preprogrammed in their acoustic structure, whereas human speech has to be learned by imitation. We know from acoustic deprivation experiments in the squirrel monkey, in which infants were raised by muted mothers, that these infants develop all fundamental call types of the species' vocal repertoire without ever having heard a call from a conspecific (Winter et al. 1974). As a consequence, a number of brain structures known to play an essential role in motor learning are dispensable for monkey calls but indispensable for human speech. Such brain structures are, for instance, the primary motor cortex, premotor cortex, caudatoputamen, cerebellum and ventrolateral thalamus. In the case of the primary motor cortex and premotor cortex, we could demonstrate that an exact reduplication of a lesion, which had caused muteness in a human patient, had no effect at all on vocalization in the squirrel monkey (Jürgens et al. 1982). In the case of the caudatoputamen, cerebellum and ventrolateral thalamus, there are several reports in the neurological literature which show that lesions within these structures cause speech disturbances in man (Botez and Barbeau 1971; Brunner et al. 1982; Darley et al. 1975; Reynolds et al. 1979; Samra et al. 1969). In contrast, lesion studies in the squirrel monkey revealed that these lesions are without effect on monkey calls (Kirzinger 1985; Kirzinger and Jürgens 1985).

A second difference between human and monkey phonation which is closely related to the first one concerns its voluntary control. While man is able to produce a wide variety of acoustic patterns, such as words, nonverbal emotional utterances and melodies, at will, vocal operant conditioning in macaques has shown that, by differentially rewarding especially long and loud calls, acoustic structure can be shaped only within very narrow limits (Sutton et al. 1973). Also, attempts to make chimpanzees utter human words failed (Kellogg 1968). Monkeys and apes thus seem to have only a very restricted voluntary control on phonation. This, however, only holds for the acoustic structure of phonations, not their initiation. Monkeys can be trained to utter a species-specific call whenever a yellow light is on and refrain from vocalizing when a blue light is on, using food as a reinforcer (Sutton et al. 1981). Monkeys thus clearly have some voluntary control on the initiation of species-specific calls but not, or only to a very limited extent, on frequency and amplitude modulation. As it is the primary motor cortex which is thought to play a crucial role in voluntary motor control in general, it does not come unexpected that the neuroanatomical projections from this area differ

1 Max-Planck-Institut für Psychiatrie, Kraepelinstr. 2, 8000 München 40, FRG

Todt/Goedeking/Symmes (Eds.)
Primate Vocal Communication

in monkey and man. While in man there is a direct connection between the primary motor cortex and the laryngeal motoneurons (Kuypers 1958), such a connection, according to our own studies, is lacking in the monkey (Jürgens 1976a).

So much about brain structures involved in human speech control but dispensable for monkey call production. What about structures relevant for monkey calls but dispensable for human speech? It is known from electrical brain stimulation studies that there are some brain structures yielding vocalization in the monkey but not in man. Such areas are the hypothalamus, amygdala, septum and midline thalamus. These areas belong to the limbic system, a group of phylogenetically old brain structures lying in the medial forebrain and thought to be involved in the control of emotional states. In fact, stimulation of these areas in man has brought about aggressive attacks, fearfulness, relaxed satisfaction as well as euphoria (Delgado 1976; Ervin et al. 1969; Heath and Mickle 1960; Sem-Jacobsen and Torkildsen 1960). In the squirrel monkey, we have tested vocalization-eliciting electrode positions in these areas for their emotion-inducing capacity (Jürgens 1976b, 1979). The animals were placed into a cage which consisted of two compartments. Presence in one compartment led automatically to stimulation of a vocalization-eliciting brain site; presence in the other compartment was devoid of stimulation. As the animal was free to move from one compartment to the other, it could switch on and off the vocalization-eliciting brain stimulation at will. The compartment in which the stimulation was given was not always the same but changed every few minutes from one side of the cage to the other. In this way, it was possible to register whether the animal avoided stimulation by always changing to the stimulation-free compartment, or followed stimulation, or remained in one compartment irrespective of the presence or absence of stimulation. The procedure thus allowed the detection of aversive and pleasurable emotional states accompanying electrically elicited vocalizations. The results of this study showed that all vocalizations elicitable from the hypothalamus, amygdala, septum and midline thalamus are accompanied by pleasurable or aversive states. Furthermore, there is a strict correlation between elicited vocalization type and accompanying emotional state in the sense that one and the same call type elicitable from different electrode sites within a specific brain area is always accompanied by the same type of self-stimulation behavior. Twitter calls, for instance, which can be obtained from the septum, are always accompanied by positive self-stimulation, that is, pleasurable emotional states; cackling calls elicitable from the amygdala and dorsal hypothalamus are accompanied by slightly aversive states, shrieking from the ventral hypothalamus and alarm peeping from the midline thalamus by highly aversive states. Furthermore, it was found that the threshold for the emotional effects was always lower than that for vocalization. These observations suggest that the vocalizations elicited from these structures do not represent primary motor responses but rather must be understood as secondary reactions due to stimulation-induced motivational effects. This points to a third difference between monkey calls and human speech, namely, the capacity of the latter to communicate information unrelated to the momentary motivational state, a capacity which seems to be lacking in monkey calls. This is not to say that monkey calls carry only information about the internal state of the vocalizer, but not about external events. The work of Seyfarth et al. (1980) on the alarm calls of green monkeys has made it clear that encoding of external events can occur to a quite specific degree. Nevertheless, these calls represent indicators of motivational states in the sense that they allow predictions about the oncoming behavior, in contrast to human speech, where verbally expressed intent and actual behavior are completely independent of each other and often enough run in opposite directions. It is, therefore, a logical consequence of the close relationship between monkey calls and motivational states that brain structures involved in motivation control also have control on monkey calls. On the other hand, we know from brain-damaged patients that lesions within the hypothalamus, amygdala, septum or midline thalamus are without effect on speech production (Darley et al. 1975; Henschen 1920; Kleist 1934).

The last group of brain structures to be discussed are those involved in monkey call as well as human speech production. Such structures, to our present knowledge, exist at three levels. The lowest is located in the lower brain stem and spinal cord and consists of the motoneurons innervating the phonatory muscles. Comparative anatomical studies have shown that the peripheral innervation of the laryngeal, expiratory and supralaryngeal muscles involved in vocal behavior is homologous in monkey and man (Beckstead and Norgren 1979; Hill 1966; Jürgens, unpublished results; Pernkopf 1952; Wyke 1974; Wyke and Kirchner 1976). Furthermore, we know from nerve transection experiments in the squirrel monkey, in which each single nerve of the three laryngeal nerves was cut unilaterally as well as bilaterally, that the functional role played by these nerves in phonation is similar in monkey and man (Jürgens et al. 1978; Thoms and Jürgens 1981). The final motor pathway for human speech and monkey calls thus seems to be the same.

The other two structures which seem to be involved in human speech as well as monkey calls are the central gray of the midbrain and the anterior limbic cortex. Both areas, when damaged in man, cause a massive reduction of spontaneous verbal utterances and, in severe cases, mutism (Barris and Schuman 1953; Botez and Carp 1968; Buge et al. 1975; Jürgens and von Cramon 1982). Those few utterances, however, which still occur are correct in articulation and grammar. So, what is affected in these cases is not motor coordination but speech impulse. In the squirrel monkey, we found that destruction of the central gray causes mutism (Jürgens and Pratt 1979). Such animals neither vocalize spontaneously nor react vocally to external stimuli, nor can they be induced to vocalize by electrical brain stimulation. Destruction of the anterior limbic cortex, on the other hand, does not mute the animals, nor does it change the acoustic structure of the calls, but severely affects voluntary vocalization control in the sense that animals are no longer able to master a vocal operant conditioning task, that is, to vocalize in order to get a food reward (Sutton et al. 1974).

The central control of monkey calls thus seems to be hierarchically organized (Fig. 1). The highest level is represented by the anterior limbic cortex, an area responsible for the vocalization of the voluntary impulse. The central gray represents a clearly lower level, as its destruction eliminates voluntary as well as unconditioned vocalization. In other words, the anterior cingulate cortex is dependent upon an intact central gray but not vice versa. The central gray seems to be an essential relay station for the initiation of all kinds of vocal behavior. It receives its input not only from the anterior cingulate cortex but also from motivation-controlling brain structures, such as the hypothalamus, amygdala, septum and midline thalamus, as well as from collaterals from primary sensory pathways. Its output goes to the lowest level within this hierarchy represented by

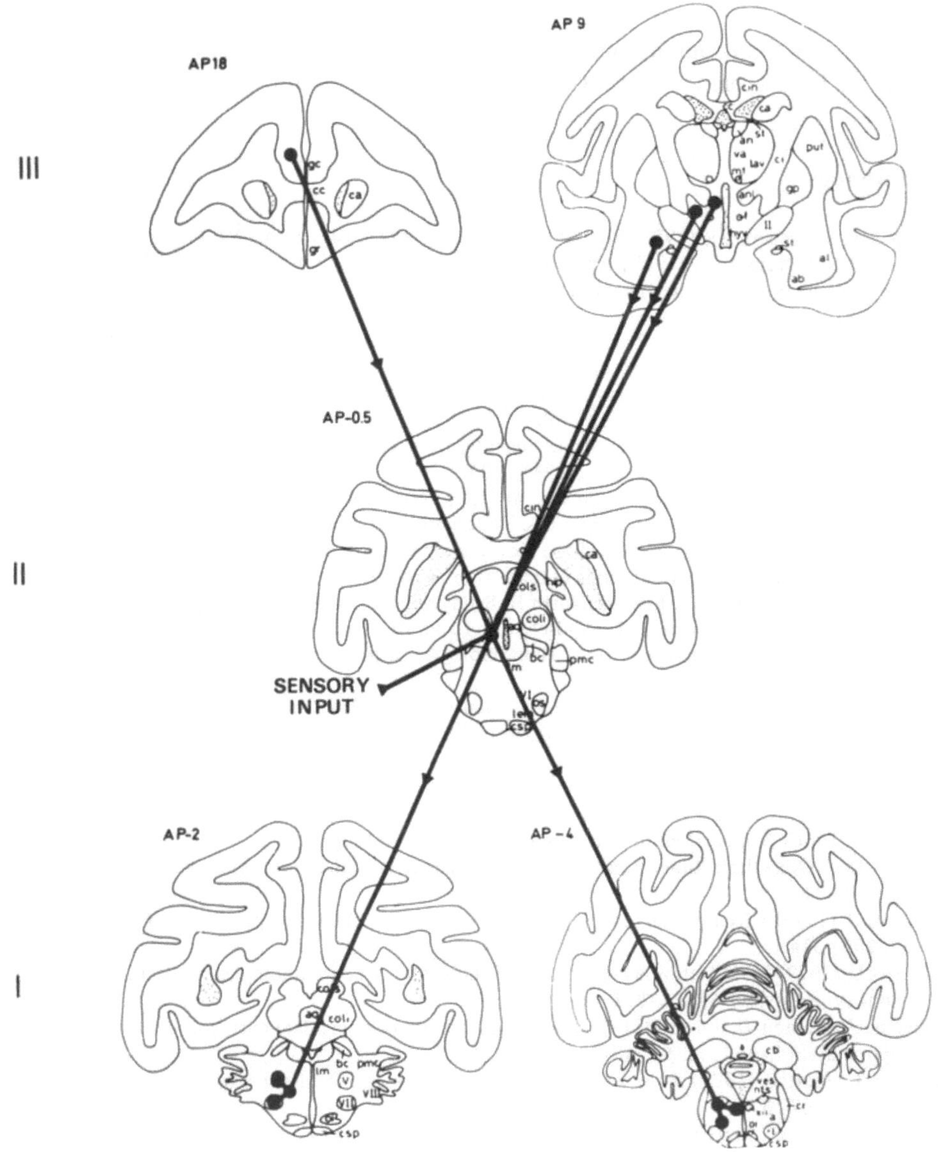

Fig. 1. Hierarchical scheme showing the central control of monkey calls. The brain diagrams at *level III* represent frontal sections of the squirrel monkey's forebrain, *level II* of the midbrain, and *level I* of the lower brain stem. The *dots* at levels *II* and *III* indicate brain areas yielding species-specific calls when electrically stimulated; the *dots* at level *I* indicate sites the stimulation of which yields abnormal phonation and fragments thereof, such as isolated vocal fold movements, lip movements or forced expiration. The *lines* interconnecting the *dots* represent anatomically verified direct connections. For further explanation, see text.

Abbreviations: *a* nucleus ambiguus; *ab* nucleus basalis amygdalae; *al* nucleus lateralis amygdalae; *an* nucleus anterior thalami; *anl* ansa lenticularis; *aq* central gray of midbrain; *bc* brachium conjunctivum; *ci* capsula interna; *cin* cingulum; *coli* colliculus inferior; *cols* colliculus superior; *cr* corpus restiforme; *csp* pyramidal tract; *f* fornix; *gc* anterior limbic cortex; *gp* globus pallidus; *gr* gyrus rectus; *hyv* ventral hypothalamus; *lav* nucleus ventralis lateralis thalami; *lem* medial lemniscus; *lm* medial longitudinal fasciculus; *mt* mammillothalamic tract; *nts* nucleus solitarius; *oi* inferior olive; *os* superior olive; *pmc* middle cerebellar peduncle; *put* putamen; *rl* nucleus reticularis lateralis; *st* stria terminalis; *va* nucleus ventralis anterior thalami; *ves* nucleus vestibulares; *II* nucleus opticus; *V* motor nucleus of trigeminus; *VI* nervus abducens; *VII* nucleus facialis; *VIII* nucleus cochlearis; *XII* nucleus hypoglossus the phonatory motoneurons and interneurons in the lower brain stem and spinal cord. This is the level where the motor coordination of vocalization takes place.

References

Barris RW, Schumann HR (1953) Bilateral anterior cingulate gyrus lesions. Neurology 3:44–52

Beckstead RM, Norgren R (1979) An autoradiographic examination of the central distribution of the trigeminal, facial, glossopharyngeal, and vagal nerves in the monkey. J Comp Neurol 184: 455–472

Botez MJ, Barbeau A (1971) Role of subcortical structures and particularly of the thalamus in mechanisms of speech and language. Int J Neurol 8:300–320

Botez MJ, Carp N (1968) Nouvelles données sur le problème du mécanisme de déclenchement de la parole. Rev Roum Neurol 5:153–158

Brunner RJ, Kornhuber HH, Seemüller E, Suger G, Wallesch CW (1982) Basal ganglia participation in language pathology. Brain Lang 16:281–299

Buge A, Escourolle R, Rancurel G, Poisson M (1975) Mutisme akinétique et ramollissement bicingulaire. 3 observations anatomo-cliniques. Rev Neurol 131:121–137

Darley FL, Aronson AE, Brown JR (1975) Motor speech disorders. Saunders, Philadelphia

Delgado JMR (1976) New orientations in brain stimulation in man. In: Wauquier A, Rolls ET (eds) Brain-stimulation reward. Elsevier/North-Holland, Amsterdam New York, pp 481–503

Ervin FR, Mark VH, Stevens J (1969) Behavioral and affective responses to brain stimulation in man. In: Zubin J, Shagass C (eds) Neurobiological aspects of psychopathology. Grune and Stratton, New York, pp 54–65

Heath RG, Mickle WA (1960) Evaluation of seven years' experience with depth electrode studies in human patients. In: Ramey ET, O'Doherty DS (eds) Electrical studies on the unanesthetized brain. Hoeber, New York, pp 214–242

Henschen SE (1920) Klinische und anatomische Beiträge zur Pathologie des Gehirns. 5. Teil: Über Aphasie, Amusie und Akalkulie. Nordiska, Stockholm

Hill WCO (1966) Primates, vol 6. Edinburgh Univ Press, Edinburgh

Jürgens U (1976a) Projections from the cortical larynx area in the squirrel monkey. Exp Brain Res 25:401–411

Jürgens U (1976b) Reinforcing concomitants of electrically elicited vocalizations. Exp Brain Res 26:203–214

Jürgens U (1979) Vocalization as an emotional indicator. A neuroethological study in the squirrel monkey. Behaviour 69:88–117

Jürgens U, Cramon D von (1982) On the role of the anterior cingulate cortex in phonation: a case report. Brain Lang 15:234–248

Jürgens U, Pratt R (1979) Role of the periaqueductal grey in vocal expression of emotion. Brain Res 167:367–378

Jürgens U, Hast M, Pratt R (1978) Effects of laryngeal nerve transection on squirrel monkey calls. J Comp Physiol 123:23–29

Jürgens U, Kirzinger A, Cramon D von (1982) The effects of deep-reaching lesions in the cortical face area on phonation. A combined case report and experimental monkey study. Cortex 18: 125–140

Kellogg WN (1968) Communication and language in the home-raised chimpanzee. Science 162: 423–427

Kirzinger A (1985) Cerebellar lesion effects on vocalization of the squirrel monkey. Behav Brain Res 16:177–181

Kirzinger A, Jürgens U (1985) The effects of brain stem lesions on vocalization in the squirrel monkey. Brain Res 358:150–162

Kleist K (1934) Gehirnpathologie. Barth, Leipzig

Kuypers HGJM (1958) Corticobulbar connexions to the pons and lower brain stem in man. Brain 81:364–388

Penfield W, Roberts L (1959) Speech and brain mechanisms. Princeton Univ Press, Princeton
Pernkopf E (1952) Topographische Anatomie des Menschen. Urban and Schwarzenberg, München
Reynolds AF, Turner PT, Harris AB, Ojemann GA, Davis LE (1979) Left thalamic hemorrhage with dysphasia: a report of five cases. Brain Lang 7:62–73
Samra J, Riklan M, Levita E, Zimmerman J, Waltz JM, Bergmann L, Coopers JS (1969) Language and speech correlates of anatomically verified lesions in thalamic surgery for parkinsonism. J Speech Hear Res 12:510–540
Sem-Jacobsen C, Torkildsen A (1960) Depth recording and electrical stimulation in the human brain. In: Ramey ER, O'Doherty DS (eds) Electrical studies on the unanesthetized brain. Hoeber, New York, pp 275–287
Seyfarth RM, Cheney DL, Marler P (1980) Monkey responses to three different alarm calls: evidence of predator classification and semantic communication. Science 210:801–803
Stevens JR, Mark VH, Ervin F, Pacheco P, Suematsu K (1969) Deep temporal stimulation in man. Arch Neurol (Chicago) 21:157–169
Sutton D, Larson C, Taylor EM, Lindeman RC (1973) Vocalization in rhesus monkeys: conditionability. Brain Res 52:225–231
Sutton D, Larson C, Lindeman RC (1974) Neocortical and limbic lesion effects on primate phonation. Brain Res 71:61–75
Sutton D, Trachy RE, Lindeman RC (1981) Vocal and nonvocal discriminative performance in monkeys. Brain Lang 14:93–105
Thoms G, Jürgens U (1981) Role of the internal laryngeal nerve in phonation: an experimental study in the squirrel monkey. Exp Neurol 74:187–203
Winter P, Handley P, Ploog D, Schott D (1974) Ontogeny of squirrel monkey calls under normal conditions and under acoustic isolation. Behaviour 47:230–239
Wyke B (1974) Ventilatory and phonatory control systems. Oxford Univ Press, Oxford New York
Wyke B, Kirchner JA (1976) Neurology of the larynx. In: Hinchcliffe R, Harrison D (eds) Scientific foundations of otolaryngology. Heinemann, London, pp 546–574

Conceptual Frameworks

Primate Vocal Expression of Affective State

KLAUS R. SCHERER[1] and ARVID KAPPAS[2]

A major feature of communication, which is all too easy to forget when one specializes in the study of particular aspects of information transmission, is the multifunctionality of signal use. Karl Bühler highlighted the fact that most signals serve several functions simultaneously in his "Organon" model of language arguing that a "sign complex" serves, at the same time, as a symbol, a semantic representation of an object or idea, as symptom, an expression of sender state, and an appeal ("Appell"), provoking a reaction of the receiver. In an attempt to classify the functions of nonverbal signs in conversation, K.R. Scherer (1980) suggested to use the semiotic approach (Morris 1946; Peirce 1931–1935) to differentiate between semantic functions (i.e., nonverbal signs replacing, amplifying, contradicting, or modifying verbal signs), pragmatic functions (expression of sender state, reactions, and intentions), and dialogic functions (relationship between sender and receiver, regulation of interaction). These types of functions can be easily mapped onto Bühler's symbol, symptom, and appeal functions, respectively. In addition, Scherer noted a syntactic function, related to the ordering of signs in a sequence as well as its hierarchical organization.

Students of communication rarely question the importance of these different functions of signal use in communication; instead they usually tend to focus on just one of these. In this chapter we will be guilty of this very problem in focussing on the expressive function of signalling, the communication of the affective or emotional state. However, we want to acknowledge that this is an analytical simplification and that one must not lose sight of the fact that most signals serve all of the functions mentioned above simultaneously. What remains to be settled, of course, is whether different aspects of the sign complexes used are specialized for certain functions. In conclusion we will return to this problem, in the attempt to suggest an alternative approach to some of the recent debates concerning the nature of primate signalling.

1 The Vocal Expression of Emotion

1.1 Expression of Emotion in Primates

Work by Eisenberg (1974, 1976), Marler and associates (Marler 1984; Green 1975; Marler and Tenaza 1977; Seyfarth and Cheney 1982), Morton (1977), Ploog and as-

1 Section de Psychologie, FPSE, Université de Genève, 1206 Genève, Suisse
2 Dept. of Psychology, Dartmouth College, Hanover, New Hampshire 03755, USA

Todt/Goedeking/Symmes (Eds.)
Primate Vocal Communication

sociates (Ploog 1974; Jürgens and Ploog 1976; Jürgens 1979, 1982), and Tembrock (1971, 1975) has produced ample evidence that most animal vocalizations express the motivational and emotional state of the organism. [However, recent work has shown that the alarm calls of vervet monkeys *(Cercopithecus aethiops)* also serve representational functions (Marler 1984; Seyfarth and Cheney 1982; see contribution by Marler and Mitani, this Vol.). We will return to this issue below.]

The importance of antecedent situations that elicit emotional reactions is mirrored in the types of categories that are commonly used to describe animal calls (alarm, threat, fear, contact, discomfort). The typical procedure consists in recording the different types of calls, noting the frequency of their occurrence, and the nature of the eliciting situations (see contribution by Dittus, this Vol.), and producing spectrographic illustrations and more or less onomatopoetic descriptions. Increasingly detailed acoustic analyses of the call are obtained. There are two major problems with this approach: (1) the difficulty of segmenting specific calls out of a stream of vocalization, and (2) the functional interpretation of the respective vocalization. In many cases the intuition of the observer enters into these decisions and it is difficult to defend functional categorizations of call systems against the reproach of subjective anthropomorphizing. The approach based on an empirical study of intraspecific responding suggested by Maurus et al. (see contribution, this Vol) provides an interesting alternative, trying to link acoustic call characteristics directly to specific response types. However, this does not quite solve the problem of determining correspondences between specific acoustic characteristics and common elements of situations that elicit specific emotions since the responses of conspecifics are determined by many different factors such as context and social position. Furthermore, it remains to be seen whether it is feasible, from the view point of research strategy, to discover the underlying relationships between acoustic dimensions and the characteristics of specific emotional states without being guided by some explicitly conceptualized theoretical assumptions. These problems are rendered even more difficult by the fact that in many primate species, calls are *graded* in terms of frequency, spectral composition and duration (Eisenberg 1974, 1976; Gould 1983; Green 1975; Jürgens 1979; Marler 1965; Marler and Tenaza 1977; Rowell and Hinde 1962; see K.R. Scherer 1985, p. 191-192, for a more detailed discussion).

Given the large number of factors that affect the acoustic characteristics of affect vocalizations in different species it may help to investigate cross-species universality and phylogenetic continuity in order to obtain insight into some basic biological patterns in affect vocalization. The idea is that the "noise" produced by species-specific variation will interfere less with the perception of the basic patterns if one looks at many species at the same time (i.e., assuming that species variations are likely to be nonsystematic and therefore cancel out).

Tembrock (1971, 1975) and Morton (1977) have investigated whether physiological factors determining the nature of phonation and resonance under emotional arousal might be similarly structured in different species.

Tembrock concludes on the basis of spectrographic analyses that there seem to be specific relationships between distinct aspects of system states and phonation characteristics resulting in certain acoustic features across different species (1975, p. 66-68, translated by KRS):

1. In the contact range, for calls reflecting states of relaxation and contentment, such as comfort and play calls, one tends to find repeated short sounds with relatively low frequencies.
2. Low frequencies also characterize dominance calls in agonistic encounters and threat calls (i.e., when critical distances are violated).
3. Defense calls, which may be the result of a transition from threat calls, are short, with a high amplitude onset and a broad frequency spectrum.
4. Submission calls, which may be the result of graded transitions from threat or defense calls, are characterized by high frequencies, repeated frequency shifts, and a tendency towards temporal prolongation.
5. Attraction calls in the distance range are also characterized by high frequencies and temporal prolongation.

Morton (1977) presents a "motivation-structural rule" concept to clarify the relation between the affective state and acoustic structure of calls: "birds and mammals use harsh, relatively low-frequency sounds when hostile and higher-frequency, more pure tonelike sounds when frightened, appeasing, or approaching in a friendly manner". The proposed selective value is that in a combat situation a low-frequency utterance might feign a large body size, whereas the high-frequency vocalization might appear more like that of an infant in order to solicit support. Morton is able to back his hypothesis by an impressive body of empirical data stemming from a large number of studies on bird and mammal sounds.

A problem with most studies of vocalization patterns is the difficulty to compare the descriptions of various call types. Tembrock and Morton had to base their comparisons on many different forms of portrayal of call type and of the underlying state of the vocalizer. Obviously the lack of consensual classification systems of acoustic characteristics and the difficulty to assess the motivational state of the sender affect the comparability of most of the studies rather negatively. Obviously it is very difficult to judge the state of an individual in free-field observation. Clearly, laboratory studies are needed to collect data suitable to a cross-individual, cross-species comparison.

In a very elegant study Jürgens (1979) induced a variety of emotional states by means of brain stimulation in squirrel monkeys *(Saimiri sciureus)* and found that the aversiveness of calls is positively correlated with total frequency range as well as with (in high-pitched harmonic calls) higher F_0 and irregularity of frequency contours. Jürgens assumes that this relationship between the acoustic structure of calls and their functional significance holds for a number of primate species and that it corresponds to the rule postulated by Morton (Jürgens 1982, p. 61).

As promising as the correspondence between the relationships reported by Tembrock, Morton, and Jürgens might appear, there are still many uncertainties and problems. Firstly, the acoustic parameters are still not very precisely defined. The question is, for example, whether "harsh" is actually equivalent to "a broad frequency spectrum". If it is, hostility should be an aversive state, like fear. Jürgens finds similar acoustic cues for several different, highly aversive states, both offensive and defensive, whereas Morton assumes harshness only for hostile states and the opposite for fear. Secondly, the specification of emotional states is still fairly imprecise. Both problems are the result of the nonexistence of conceptual and operational description systems for characteristics of emotional states or acoustic features of complex vocalizations (a more detailed review of this issue can be found in K.R. Scherer 1985, pp. 191–195).

1.2 Expression of Emotion in Man

Though Charles Darwin (1872) pointed out quite early the importance of voice characteristics for the expression of emotions, the human voice has been but recently a subject for thorough analysis by psychologists. This might be a result of the very nature of speech itself: It is very volatile. Whereas the face could be depicted very early in paintings or drawings, voice characteristics had to be recorded on magnetic audiotape and analyzed by means of sophisticated equipment like spectrographs and more recently digital analysis.

Though methods improved over the last decades it proved to be difficult to correlate acoustic features of vocal utterances with affective states. Two procedures have been typically used: The first consists of inducing emotional states in human subjects or to ask actors to portray such states and to consequently measure a set of acoustic parameters like mean or standard deviation of F_0 in the vocalizations produced in order to detect affect-specific vocal patterns. These experiments aim at the study of the vocal *expression* of emotion. The second procedure uses judgement studies in which vocal stimuli (usually vocal emotion portrayals by actors) are presented to naive subjects who are asked to attribute the emotional state of the sender. Their ratings on various emotion scales are then correlated with acoustic parameters. These experiements aim at the *impression* aspect of the communication process. The stimuli were either taken from natural interactions or staged interactions in the laboratory.

Table 1 presents a summary of the major findings on the vocal expression of human emotions, including both common emotion labels and major dimensions of feeling. It contains specifications of the nature of particular cues for an emotion only in those cases, where independent empirical studies yielded the same outcome. The question

Table 1. Summary of results on vocal indicators of emotional states[a]

Emotion	Pitch Level	Pitch Range	Pitch Variability	Loudness	Tempo
Happiness/joy	High	?	Large	Loud	Fast
Confidence	High	?	?	Loud	Fast
Anger	High	Wide	Large	Loud	Fast
Fear	High	Wide	Large	?	Fast
Indifference	Low	Narrow	Small	?	Fast
Contempt	Low	Wide	?	Loud	Slow
Boredom	Low	Narrow	?	Soft	Slow
Grief/sadness	Low	Narrow	Small	Soft	Slow
Evaluation	?	?	?	Loud	?
Activation	High	Wide	?	Loud	Fast
Potency	?	?	?	Loud	?

[a] From *Speech evaluation in psychiatry* (p. 206) by J. Darby (Ed.), 1981, New York: Grune & Stratton. Copyright 1981 by K.R. Scherer. Reprinted by permission.

marks in some cells indicate a lack of appropriate studies or disagreement in the findings. The results indicate a differentiation for high activity emotions (joy, anger, and fear) and low activity emotions (sadness, boredom). The former seem to be characterized by high frequency, intensity, and tempo, the latter by low frequency, intensity, and tempo. One conclusion from these somewhat disappointing outcomes might be that vocal parameters transmit only information about the level of physiological arousal of an individual. However, this seems very unlikely. Everyday experience and many empirical studies show that different emotions can be very well identified and distinguished from each other by means of vocal cues alone. Obviously we are confronted by a paradox: on the one hand, there are dozens of independent experiments unable to find typical acoustic correlates of specific emotional states, on the other hand, there is an equal number of studies showing that the recognition of emotions from voice and speech samples is as good or better than from facial expressions (average recognition rate of 56% accuracy, compared with 12% expected by chance, in 28 studies; see K.R. Scherer 1986).

Another possible conclusion is that the voice does carry sufficient information to differentiate various emotions, but that the relevant acoustic cues have not been identified and examined so far. The acoustic parameters that have been investigated in the past, mostly energy and fundamental frequency, may very well be mainly related to physiological arousal. Others, like voice quality, i.e., spectral parameters, might be better indicators of specific affect states. There is some evidence that this might indeed be the case. Williams and Stevens (1972) could identify a number of additional characteristics that seem to differ for the emotions under study. These included irregularity of phonation, fundamental frequency contours, and energy distribution in the spectrum. Judgment studies using synthesized stimuli have shown that spectral energy distribution changes produce differential impressions of specific emotions, for example, anger is more likely to be attributed for stimuli with greater high frequency energy. Thus there is a strong possibility that specific acoustic characteristics do exist for different emotional states. Unfortunately, our current state of knowledge is not yet sufficiently advanced to allow their definite identification.

2 A Theoretical Model of Vocal Affect Expression

One of the most serious problems in this area of research has been the lack of theoretical models that could serve to generate hypotheses for systematic empirical study. Recently, K.R. Scherer (1986) has used his component process model of emotion to generate a number of specific predictions on affect-induced vocalization changes. Details of this model can be found in K.R. Scherer (1985, 1986). The following section will provide only a rough summary of the argument.

The basic assumption of the model is that the evaluation or appraisal of a situation or event which produces an emotional state consists of a sequential process of five stimulus evaluation checks (SECs). These checks (and their more detailed subchecks) are defined as follows:

1. *Novelty check.* Evaluation of whether there is a change in the pattern of external or internal stimulation, particularly whether a novel event occurred or is to be expected.

2. *Intrinsic pleasantness check.* Evaluation of whether a stimulus event is pleasant, inducing approach tendencies, or unpleasant, inducing avoidance tendencies; based on innate feature detectors or on learned associations.
3. *Goal/need significance check.* Evaluation of whether a stimulus event is relevant to important goals or needs of the organism (relevance subcheck), whether the outcome is consistent with or discrepant from the state expected for this point in the goal/plan sequence (expectation subcheck), whether it is conducive or obstructive to reaching the respective goals or satisfying the relevant needs (conduciveness check), and how urgently some kind of behavioral response is required (urgency subcheck).
4. *Coping potential check.* Evaluation of the causation of a stimulus event (causation subcheck) and the coping potential available to the organism, particularly the degree of control over the event or its consequences (control subcheck), the relative power of the organism to change or avoid the outcome through fight or flight (power subcheck), and the potential for adjustment to the final outcome via internal restructuring (adjustment subcheck).
5. *Norm/self-compatibility check.* Evaluation of whether the event, particularly an action, conforms to social norms, cultural conventions, or expectations of significant others (external standards subcheck), and whether it is consistent with internalized norms or standards as part of the self-concept or ideal self (internal standards subcheck).

The differentiation of the emotions is considered to be determined by the specific combination of outcomes of these checks. For example, anger is suggested to result when an event is evaluated as highly relevant to the actual goals/needs of the organism, discrepant from expectation, hindering goal achievement but where the organism assumes to have enough control and power to avoid negative outcomes. For fear, on the other hand, the coping potential evaluation would signal a lack of power to successfully influence the event. Table 2 shows a hypothetical prediction table for the check results which are expected to produce particular emotions.

This model departs from other theories of emotion in that the responses are considered to be determined by the outcomes of the individual checks rather than by a central mechanism specific to an emotion as a whole. In other words, each individual check outcome is expected to modify the state of the various organismic subsystems: the central nervous system (CNS), the autonomic nervous system (ANS), and the somatic system (in particular, the striped musculature). Detailed predictions on the "component patterning" of these systems by the check outcomes can be found in K.R. Scherer (1985, 1986). In the present context, our main interest is the effect of different check outcomes on respiration, phonation, and articulation processes. The respective predictions are reproduced in Table 3.

By combining the information in Tables 2 and 3 we can arrive at specific hypotheses concerning the acoustic parameters characterizing vocalization under specific emotional states. These predictions are listed in Table 4.

Obviously, much of the preceding is highly speculative in nature. Yet, it would seem that relatively concise hypotheses might help to focus research on vocal affect expression in such a way that comparability of research results becomes possible allowing a greater convergence of research effort. Whether the predictions ventured above can be maintained depends of course on the results of empirical work to be done in an

Table 2. Hypothetical outcomes of stimulus evaluation checks for selected emotional states (K.R. Scherer 1986, p. 147)

Emotional state	Novelty	Pleasantness	Goal/need significance				Coping potential			Norm compatibility	
			Relevance	Expectation	Conduciveness	Urgency	Control	Power	Adjust	External	Internal
Enjoyment/happiness	Low	High	Medium	Consistent	High	Very low	–	–	High	High	High
Elation/joy	High	High	High	Discrepant	High	Low	–	–	Medium	High	High
Displeasure/disgust	Open	Very low	Low	Discrepant	Low	Medium	Open	Open	High	Low	–
Contempt/scorn	Open	Low	Low	Discrepant	Low	Low	Open	High	High	Low	–
Sadness/dejection	Low	Low	High	Discrepant	Obstruct	Low	None	–	Medium	–	–
Grief/desperation	High	Low	High	Discrepant	Obstruct	High	Low	Low	Low	–	–
Anxiety/worry	Low	Open	Medium	Discrepant	Obstruct	Medium	Open	Low	Medium	–	–
Fear/terror	High	Low	High	Discrepant	Obstruct	Very high	Open	Very low	Medium	–	–
Irritation/cold anger	Low	Open	Medium	Discrepant	Obstruct	Medium	High	Medium	High	Low	Low
Rage/hot anger	High	Open	High	Discrepant	Obstruct	High	High	High	High	Low	Low
Boredom/indifference	Very low	Open	Low	Consistent	Obstruct	Low	Medium	Medium	High	–	–
Shame/guilt	Low	Open	High	Discrepant	Obstruct	Medium	High	Open	Medium	Very low	Very low

Table 3. Component patterning theory predictions of vocal changes after different SEC outcomes (K.R. Scherer 1986, p. 156)

Novelty check	
Novel	Old
Interruption of phonation Sudden inhalation Silence Ingressive (fricative) sound with a glottal stop (noise like spectrum)	No change
Intrinsic pleasantness check	
Pleasant	Unpleasant
Faucal and pharyngeal expansion, relaxation of tract walls	Faucal and pharyngeal constriction, tensing of tract walls
Vocal tract shortened by mouth corners retracted upward	Vocal tract shortened by mouth corners retracted downward
More low-frequency energy. F1 falling, slightly broader F1 bandwidth, velopharyngeal nasality	More high-frequency energy. F1 rising, F2 and F3 falling, narrow F1 bandwidth, laryngopharyngeal nasality
Resonances raised	Resonances raised
Wide voice	*Narrow voice*
Goal/need significance check	
Relevant and consistent	Relevant and discrepant
Shift toward trophotropic side: overall relaxation of vocal apparatus, increased salivation	Ergotropic dominance: overall tensing of vocal apparatus and respiratory system, decreased salivation
F0 at lower end of range, low to moderate amplitude, balanced resonance with slight decrease in high-frequency energy	F0 and amplitude increase, jitter and shimmer, increase in high-frequency energy, narrow F1 bandwidth, pronounced formant frequency differences
Relaxed voice	*Tense voice*
If event conducive to goal: relaxed voice + wide voice	If event conducive to goal: tense voice + wide voice
If event obstructive to goal: relaxed voice + narrow voice	If event obstructive to goal: tense voice + narrow voice
Coping potential check	
Control	No control
Ergotropic dominance: (see tense voice)	Trophotropic dominance: Hypotension of the musculature in the vocal apparatus and respiratory system
(See tense voice)	Low F0 and restricted F0 range, low-amplitude, weak pulses, very low high-frequency energy, spectral noise, formant frequencies tending toward neutral setting, broad F1 bandwidth
Tense voice	*Lax voice*

Table 3 (continued)

Coping potential check	
Power	No power
Deep, forceful respiration; chest register phonation Low F0, high-amplitude, strong energy in entire frequency range *Full voice*	Rapid, shallow respiration; head register phonation Raised F0, widely spaced harmonics with relatively low energy *Thin voice*
Norm/self-compatibility check	
Standards surpassed	Standards violated
Wide voice + full voice + Relaxed voice (if expected) + Tense voice (if unexpected)	Narrow voice + thin voice + Lax voice (if no control) + Tense voice (if control)

attempt to test the predictions. Similarly, the utility of the component process model and the patterning assumption depends on empirical verification.

2.1 Push and Pull Effects in Vocal Affect Expression

Most of the preceding discussion assumed that vocal features are directly determined by physiological changes in the organisms that accompany emotional arousal. For example, muslce tension will change in the direction of facilitating adaptive action in response to an external event and phonation is expected to law-fully change because of the type of muscle potential change that ensues. One can describe this in terms of the physiology "pushing" phonation. However, it is unlikely that all vocalization parameters are exclusively determined by this "internal push". We have argued that other factors, particularly social-conventional factors, are likely to exert an "external pull" on vocalization, resulting in changes that may at times be at odds with physiological effects (see U. Scherer et al. 1980). In this section we will discuss in which ways the predictions ventured above need to be tempered by including possible external pull effects.

2.2 Defining "Push" vs "Pull"

It ssems useful to preface this discussion with an overview of the major factors "pushing" the vocal apparatus as well as the different "pull" factors that can modify vocal expression. It is appropriate to first define these two types of determinants in a more stringent fashion.

Push factors are defined as changes in states of subsystems in the organism which have a direct effect on vocalization parameters. Push factors work in a largely involuntary fashion, their effects on vocal settings and the resulting acoustic parameters are almost exclusively determined by the nature and force of the physiological mechanisms involved. Pull factors, on the other hand, although mediated through internal systems

Table 4. Changes predicted for selected acoustic parameters under different emotional states

	Emotions[a]											
Acoustic parameters	ENJ/ HAP	ELA/ JOY	DISP/ DISG	CON/ SCO	SAD/ DEJ	GRI/ DES	ANX/ WOR	FEAR/ TER	IRR/ COA	RAGE/ HOA	BOR/ IND	SHA/ GUI
F0												
Perturbation	<=	>			>	>		≥		>		
Mean	<	≥	>	><	<>	>	>	≥≥	><	><	≤	>
Range	<=	≥			≤	>		≥≥	<	≥		
Variability	<	≥			≤	>		≥≥	<	≥		
Contour	<	>			≤	>	>	>>	<	=		>
Shift regularity	=	<						<		<	>	
F1 mean	<	<	>	>	>	>	>	>	>	>	>	>
F2 mean			<	<	<	<	<	<	<	<	<	<
F1 bandwidth	>	><	<<	<	<>	<<	<	<<	<<	<<	<	<
Formant precision		>	>	>	≤	>	>	>	>	>		>
Intensity												
Mean	<	≥	>	>>	≤≤	>		>	≥	≥≥	<>	
Range	<=	>			<			>	>	>		
Variability	<	>			<			>		>		
Frequency range	>	>	>	>>	>	>>		>>	>	>	>	
High-frequency energy	<	<>	>	>	><	>>	>	≥≥	>>	>>	><	>
Spectral noise					>							
Speech rate	<	≥			≤	>		≥≥		≥		
Transition time	>	<			>	<		<		<		

[a] ANX/WOR = anxiety/worry; BOR/IND = boredom/indifference; CON/SCO = contempt/scorn; DISP/DISG = displeasure/disgust; ELA/JOY = elation/joy; ENJ/HAP = enjoyment/happiness; FEAR/TER = fear/terror; GRI/DES = grief/desperation; IRR/COA = irritation/cold anger; RAGE/HOA = rage/hot anger; SAD/DEJ = sadness/dejection; SHA/GUI = shame/guilt. F0 = fundamental frequency. F1 = first formant. F2 = second formant. > = increase. < = decrease. Double symbols indicate increased predicted strength of the change. Two symbols pointing in opposite directions refer to cases in which antecedent voice types exert opposing influences (K.R. Scherer 1986, p. 158).

(e.g., the monitor and executive systems) are externally based in that they operate towards the production of specific acoustic patterns with optimal transmission features or socially defined signal values. Another way to characterize the distinction is the following: For push factors, the mechanism is the independent variable (the physiological processes affecting vocalization) and the acoustic result is the dependent variable, whereas for pull factors, the acoustic pattern to result from vocalization is the independent, the mechanism to achieve this the dependent variable. The following example (from K.R. Scherer 1985, p. 209) may be helpful: "If an object is pushed down a hillside, the "pusher" usually does not intend to determine or predict the exact location of the final resting point, yet the laws of physics determine the course and the final destination of the object. However, if an object is pulled up a hillside, the "puller" usually has very definite intentions concerning the final resting point of the object and will attempt to overcome all those physical forces obstructing the desired course."

Most of the pull effects are likely to involve the somatic nervous system, since it is unlikely that the ANS can be fine-tuned by central processes to provide the conditions for the production of particular acoustic characteristics (e.g., by the proper dosage of salivation). Push effects, on the other hand, are based on both the ANS and the somatic nervous system. Finally, pull effects in another expressive modality can serve as push factors for vocalization. For example, the deceptive use of a bulging chest or a particularly grim threat face with a screwed-up mouth will strongly influence vocalization, without a specific acoustic pattern being aimed at by the signaller. However, in most cases this distinction will be difficult to make since it is highly likely that in deceptive signalling (see below) all of the expressive features, including the vocal ones, are produced by the respective pull factors.

2.3 Types of Pull Effects

As defined above, pull factors are characterized by the fact that a specific acoustic pattern is to be produced for communicative purposes. A preliminary typology of pull effects can be attempted on the basis of the requirements or the functions of the acoustic pattern "pulled" in the respective case.

Transmission and Localization. An important aspect of communicatively used signals concerns their transmission characteristics for sending over a distance and their utility as cues for the sender's location.

Depending on the habitat of a species, or specific situational requiremens, vocal characteristics have to be chosen by the sender in such a way as to ensure that the message reaches the receiver.

Similarly, the optimal acoustic pattern differs depending on whether the sender wants to draw attention to himself or attract a conspecific, or whether, as in flight, a warning of other group members without giving away the location of the sender is adaptive (see the contribution by Waser and Brown, this Vol.).

Self-Presentation. This is one of the most important, and most controversial, pull factors. As shown above, in many species, agonistic encounters with threat displays by the opponents are characterized by the use of low pitch (Jürgens 1979; Morton 1977; Tembrock 1975). It is interesting to speculate to what extent these correspondences are brought about by pull or push effects.

Morton attributed these relationships to an "ancient coupling" between body size and sound frequency (i.e., the bigger, more powerful, and thus confidently hostile animal producing low-frequency harsh phonation, i.e., a push effect). It is reasonable to assume that selective pressure works in the direction of low-frequency phonation to appear "big" to the opponent and gain an advantage in the, often ritualized, fights. However, Morton argued that this signal has evolved in such a way as to be determined almost exclusively by pull effects:

"The argument is that genes that promote fighting, where large body size is favored, will tend to be replaced by genes involved in the communication of motivation if the communication genes are as effective as fighting in acquiring the resources in question." (Morton 1977, p. 865).

Recent sociobiological arguments (Dawkins and Krebs 1978; Caryl 1979) push this notion further, assuming that virtually all communicative signals are deceptive and manipulative, determined by strategic intention rather than underlying motivational state. In other words, according to this model, most vocalization patterns would be determined by pull effects.

This view has been challenged by a number of ethologists (Hinde 1981; Zahavi 1982). On the basis of these and similar arguments, K.R. Scherer (1985) has attempted to show that this extreme view implies that signals would become unreliable and worthless in the long run, arguing that deceptive signal use, indicating nothing but the strategic intent of the signaller, can only work some of the time, under the condition that signals provide more reliable information about the actual presence of the referent most of the time. In other words, if signals carried no other information than the deceptive intent of the sender, they should quickly lose their adaptive value.

Another important point is the signal use by the receiver, particularly in terms of the trust placed in the veridicality of the sender's signalling. Goffman (1969) has very elegantly described the complexities in the deceptive signal arms race. As might be expected, a signal is seen as all the more trustworthy the closer it comes to being determined by push effects. K.R. Scherer (1977) has argued that "raw" affect vocalizations will be seen as more spontaneous and reliable signals than conventionalized, ritualized "vocal emblems". This analysis leads to the conclusion that, in general, pull effects are likely to remain closely linked to the push effects upon which they are based (see detailed discussion in K.R. Scherer 1985, p. 212–215).

This controversy highlights two important points: (1) for deceptive signals to work, there must be a valid underlying relationship between signal and referent; in other words, self-presentation pull effects can only work if there are corresponding push effects, and (2) deceptive signal use must be limited in degree and frequency of usage in order not to lose its basis, the overall veridicality of the information transmitted by social signals.

The importance of self-presentation has been an important topic in social psychology long before the current debate in sociobiology, with one of the most important contributions coming from E. Goffman (1959). In this tradition, the assumption is that the purpose of controlling signals was not one of outright deception, but more of ameliorative correction in the presentation of oneself towards others. Thus, in line with the view of self-presentation in the social-psychological literature (Goffman 1969; Schlenker 1980), it is expected that the individual, in signalling, will try to enhance positive information about itself (power, ability, likability) and to play down negative information, but that these attempts do not normally lead to the disappearance of all correlation between signal and underlying states and traits. Although this does occur – after all, impostors do exist in animals and humans – it tends to be infrequent because of the high cost and risk involved (see K.R. Scherer 1985). The important corrective, for safeguarding the requirements for deceptive signalling mentioned above, is the possibility that often the bluff can be called at any point. This will effectively limit the extent of mismatch between signal and underlying state and the number of situations in which deceptive signalling is used strategically.

The view taken here, then, is that pull effects are common in the signalling of power and confidence but that they have not completely replaced the push effects; rather they tend to interact with each other in determining the acoustic features of the vocalization.

Attracting or Repelling Conspecifics. Again, Darwin was one of the first to mention the importance of this pull factor:

"When male animals utter sounds in order to please the females, they would naturally employ those which are sweet to the ears of the species; and it appears that the same sounds are often pleasing to widely different animals, owing to the similarity of their nervous systems ... On the other hand, sounds produced in order to strike terror into an enemy, would naturally be harsh and displeasing." (1872/1965, p. 91).

Thus, the essential determinant for the kind of acoustic pattern required here is the pleasantness or unpleasantness of this pattern (and implicitly, the approach or avoidance tendencies generated by it) to the group of intended receivers.

Accommodation. An intriguing pull effect repeatedly demonstrated by psychologists studying language in social interaction is accommodation or adaptation of various features of one's vocal behavior to that of interaction partners with whom one has established a positive relationship (Feldstein and Welkowitz 1978; Giles and Smith 1979). Here, the acoustic pattern produced by the partner is the template to be matched, the result towards which the vocal settings are pulled.

Conventional Social Signals. There are a number of vocalizations with specific acoustic features that are conventionalized in a culture and that have shared meaning. These are stylized interjections or vocal emblems (see Goffman 1979; K.R. Scherer 1977) and specific vocal styles or situational registers (see P. Brown and Levinson 1979; Laver and Trudgill 1979). In this case, the acoustic patterns to be produced are socially prescribed.

The last three types of pull effects have been described only briefly since they seem to play a less important role in the vocal communication of emotion than the first two types. The coverage of these phenomena in the present context cannot do justice to their general importance in vocal interaction.

Although the respective predominance of push and pull effects will vary with the situation, particularly with the level of need for strategic signalling, both would tend to affect the acoustic patterns of vocalization in the same direction (although the underlying state signalled will not be present if pull effects dominate). The multifunctionality of the same acoustic factors for both push and pull effects can also be shown in cases where the pull effects may not be originally based on push effects. For example, the source of harsh sounds seems more easily locatable than that of tonelike, harmonic sounds, which is obviously in the interest of the fighter, waiting for his opponent, and the one who flees, hoping to go unnoticed, respectively. C. Brown (1982) concludes a review of the literature in this area as follows:

"Collectively, these observations are consistent with the proposition that the level of arousal and the level of contact seeking registered by different calls may be communicatively reemphasized by the signal's relative locatability. The ease of localization may be one of a number of prelinguistic codes in which a facet of the affective state of the vocalizer covaries with a perceptual dimension." (C. Brown 1982, p. 159).

Following the discussion of the push factors, some of the potential pull factors connected with the respective SEC outcome will be discussed. There are two major issues of interest here. Does the outcome of the SEC generate communication pressure rendering it likely: (1) that pull effects on vocalization resemble ritualized or conventionalized push effects, resulting in stereotyped signals that can be used in deceptive signalling, and (2) does the strategic use of signals indicating (or feigning) a particular SEC outcome (including the behavioral intentions generated) require the production of acoustic features that are *not* part of the push effects, e.g., features involving transmission characteristics, pleasantness of sound, iconic representation of a referent, or arbitrary social conventions concerning signal structure. Although in the case of conventionalized push effects pull factors would tend to enhance and possibly exaggerate some of the push effects to increase signal clarity, in the case of discrepant pull factors these could enhance, add to, or counteract the push effects. Given the complexity of these processes and the lack of detailed knowledge about the evolutionary development of vocal signalling, only a few speculative observations can be offered.

2.4 Potential Pull Effects for Individual Checks

Novelty Check. In terms of self-presentation it does not seem very adaptive to produce any auditory signal, allowing localization of an individual, before the potential dangerousness of a novel stimulus has been established. There could be adaptive value in the production of a social signal communicating the presence of a novel stimulus to the group but the potential cost of many false alarms may outweigh these advantages. Again, further processing of the novel stimulus event may be required before the onset of any vocalization.

Intrinsic Pleasantness Check. In terms of the communicative function of emotion, it seems reasonable to assume that there is adaptive value in letting other group members know that a particular object has produced approach or avoidance tendencies, reducing the overall amount of trial-and-error behavior and its costs. Thus, any expressive intention movement indicating approach should produce pleasant associations in others, whereas avoidance signals are likely to have unpleasant connotations.

The activity of the facial muscles, particularly in the mouth region, offers a fascinating topic for the discussion of pull effects in relation to push effects. If one assumes that the human smile evolved from primate "play faces" or "relaxed open mouth faces" (van Hoff 1972; see also Redican 1982, p. 253–258) or as a residue of the response to sweet tastes (Bühler 1933), the resonance changes (formants moving up) brought about by the shortening of the vocal tract would have to be considered push effects.

Ohala (1980) has proposed the interesting theory that the open-mouthed smile accompanied by mouth corner retraction is a *consequence* of adaptive pressures to produce high-pitched, tonelike sounds in friendly or submissive encounters (based on Morton 1977), with the required acoustic pattern as a pull factor. However, Ekman (personal communication) points out that a variety of different muscles with differential effects on the shape of the oral part of the vocal tract, many of which are not necessarily associated with smiling, could be involved here. Given our incomplete knowledge about the effect of the different muscular actions that affect mouth shape and the movement of the corners of the mouth (see Ekman and Friesen 1978) on the resonance characteristics of the vocal tract, it is not possible at present to evaluate the relative merit of the different theories on the origin of the smile.

Although this is not the place to discuss the dual factor theory proposed above in relation to facial expression, it would seem that the discussion on the origin of facial displays (see the excellent summary in Redican 1982) may benefit from considering the possibility that both push and pull factors may be involved simultaneously in shaping facial expression. For example, malicious smiles are likely to consist of elements of both aggression and enjoyment (two push factors). Submissive smiling, on the other hand, may be composed of a push component, a fear display, and a pull component consisting of attempts to mask the fear response and, possibly, the strategic use of an enjoyment smile to induce positive mood.

Goal/Need Significance Check. The major pull factor for this check would seem to be self-presentation as frustrated or content, as the case may be, to communicate the reaction to an outcome of a stimulus event. In terms of the signalling of behavioral intention, the vocalization could function in such a way as to indicate the high activity level of the organism under ergotropically aroused states, possibly to alert conspecifics to outcomes discrepant from expectation. The acoustic features produced by the push effects seem well suited for this purpose and it seems unlikely that there are acoustic target patterns originally induced by these pull factors.

Coping Potential Check. The signalling of power and confidence, a result of a positive outcome of this check, is ripe with pull effects. As the ethological literature shows, in many species, including *Homo sapiens*, most agonistic encounters are settled by ritualized exchanges of intention movements, in this case mostly threat displays, rather than actual fights. In a large number of different species, vocalization seems to play a major role in these encounters. The evidence reviewed above, showing that generally that birds and mammals use harsh, relatively low-frequency sounds when hostile and higher-frequency, more pure tonelike sounds when frightened, appeasing, or approaching in a friendly manner (e.g., Jürgens 1982; Morton 1977; Tembrock 1975; see review in K.R. Scherer 1985) underlines the role of vocal communication in this context.

In line with the nature of the push effects, one would expect self-presentation pull effects for the confident power subcheck outcome to consist of strong vocal effort and chest register. Pull effects should work in the direction of deemphasizing tense voice since this indicates ergotropic arousal. As shown above, this type of arousal should be fairly low in genuinely confident and powerful individuals and self-presentation should aim to mask any traces of tenseness or harshness.

An observation reported by Zahavi (1982) is interesting in this context:

"My experience with human vocalizations, during aggressive encounters, suggests that the same individual threatens with a relaxed vocal signal of a low pitch when confronting an individual which is inferior to him in his fighting potential and raises the pitch of his voice when confronting a superior fighter. An inferior fighter is more likely to lose a fight if it relaxes in the face of its opponent. Hence cheating by the use of a relaxed vocal signal demands a high cost from the signaler." (p. 6).

Morton's (1977) review finding harshness in hostile encounters for many species, which seems to contradict this view, may reflect the fact that in most agonistic encounters there is not just a strategic display of power and confidence but also a high degree of ergotropic arousal given the uncertainty of the power check and the likely outcome of a fight. As Hinde (1981) points out, most threat displays reflect a motive conflict between attack and flight, which in terms of the componential patterning model, should result in elements of both chest voice and tense voice. Furthermore, harshness is much more audible in the lower F0 regions (Laver 1980, p. 127–128). Since most dominant animals are likely to vocalize at lower fundamental frequency this may result in stronger harshness impressions without necessarily signalling greater ergotropic arousal.

Similarly, the pull effects described by Morton for "frightened, appeasing, or friendly" impressions may consist of blends based on the outcomes of several SECs and the resulting push effects. The "higher-frequency tonelike sounds" described by Morton may be produced in different ways depending on the relevance of particular SEC outcomes and the communicative pressure generated by them. Thus, in terms of the model and the predictions presented above, submission and appeasement, i.e., a nonconfident outcome of the power check, should lead to the use of a thin voice but not necessarily a tense voice. Fright should be best communicated by a tense voice plus a thin voice. Friendliness, on the other hand, should involve a positive outcome of the pleasantness check and be best represented by a wide voice plus a thin voice (in order not to invoke a status or power differential). These predictions are explored in more detail in K.R. Scherer (1985).

The pull effects for a low or no control outcome are difficult to predict. In terms of the adaptive function, the need of the organism to adjust to external events through internal restructuring, it seems reasonable to postulate a signal to the social surround that the individual does not want to be disturbed and focusses inward. The passivity implied by the lax voice seems well suited for this purpose. A different situation is presented by the situational requirements of seeking help or assistance from others in a condition of helplessness, which may involve a combination of a narrow voice, lax voice, and thin voice. This is different of course from *active supplication* of help which would seem to require a narrow voice, thin voice, and *tense* voice. A detailed discussion of the pull effects expected in this condition (both in the preceding section on push and pull effects and in this section) shows that they are likely to have evolved on the basis of push effects and work in the same direction. Because of the strategic importance of the power dimension in social organization and survival, these pull effects can be expected to frequently operate in addition to, or in some cases even in the absence of push effects. However, it seems unlikely that they have totally replaced the push effects.

Norm/Self-Compatibility Check. Very little was said about the effects of the outcomes of this check when the push effects were discussed above. The reason is that we know very little about the physiological and communicative correlates of this dimension. Pull effects are equally difficult to predict. One might speculate on self ideal congruence/ discrepancy effects: The general assumption is that shame (self ideal incompatibility) leads to avoidance of social contacts, shunning the attention from others whereas pride (self ideal congruence) should have the opposite effect, seeking social attention and recognition. Consequently, one could argue that any vocal parameters likely to arouse attention by others would be attenuated or amplified respectively.

3 Affect Expression in the Context of Social Markers

So far, we have discussed the effects that internal push factors (mostly of a physiological nature) and external pull factors (mostly related to social convention and environmental constraints) have on affect vocalizations. This discussion may have sounded as though these effects are essentially equivalent for all individuals across all species. This is of course not the case. While we do believe that the predictions developed above are based on some very basic principles for which we have claimed phylogenetic continuity and some degree of universality, we certainly acknowledge the existence of species-specific factors and individual differences. Indeed, it is most likely that affect expression is intricately bound up with what we have called "social markers" in vocalization (K.R. Scherer and Giles 1979). These markers are seen to provide the receiver with information about the sender's biological, social, and psychological characteristics. As soon as social organization is based on enduring relationships between individuals, such as in stable mating pairs or in dominance hierarchies, the identifiability of individuals becomes essential.

The marking of individual identity, categorical membership, and other characteristics of the sender is not only an incidental, unavoidable by-product of any act of communication, it is necessary for the receiver to respond to the message conveyed by the sender since the interpretation of the message often requires knowledge of the sender's identity.

The particular signals and the sensory channels that are used for marking are determined by the communication systems that are mainly used by a species. Marking may occur by means of visual, auditory, or chemical communication. Examples for the marking of biological, psychological, and social categories have been shown for many subhuman species as well as man.

Since many of the social markers that use the vocal communication channel affect the same acoustic parameters as the affect expression push and pull effects, one can expect a very complex interaction of determinants. In studying vocal affect expression the potential effects of social markers have to be kept in mind.

It may be useful at this point to briefly discuss the types of relationships between vocal parameters and the states or referents they signal. Giles et al. (1979) have suggested to distinguish between three dimensions that characterize the coding form for the signal. The first dimension is "discrete vs continuous". Markers that are discrete can be described as "the sender *has it*" vs "the sender *has it not*". Continuous vocal markers vary to the degree that the marked variable changes (for example, the correlation of vocal tract length and fundamental frequency). Social, biological, and psychological states may also be either discrete or continuous. Sex, for example, is discrete, whereas arousal is usually rather continuous. A second important aspect of marking is the coding strength, i.e., the nature of the relationship between markers and marked state in terms of strength of their correspondence. Two sets of distinctions are important in this respect. The first relates the proportion of marker occurrences and marked states. The proportion may be *invariant, conditional,* or *probabilistic*. A vocal marker is invariant, when it is always associated with a marked variable. Conditional vocal markers are those, where the relationship between vocal feature and marked state appears only in specific contexts, situations, or with certain types of individuals. An example might be a certain type of alarm call that is only found in specific threatening situations and not others. Most common are probabilistic markers that mark a particular state for some percentage of time, for some percentage of individuals, without actual determinants of its occurrence being known. If more information is known about the determinants of probabilistic vocal markers to an extent where it is possible to specify conditions of their occurrence, these can then be assumed to be actually conditional vocal markers.

The second distinction concerns the covariation between a vocal feature and a marked variable. If the former marks one (and exclusively one) state it is called a unique vocal marker. If instead two or even more states are designated by one marker it may be called equivocal.

So far we have examined the vocal signalling of affect almost exclusively from the sender perspective. Yet, another important aspect in the investigation of vocal signalling is the consideration of the receivers perspectives. If there is a strong correlation between a vocal marker and a marked state but the marker is not being processed by the receiver, the marker is called *etic*. An etic marker is a potential source of information that has not been exploited by the interaction partner. If the receiver does use this information, the marker is called *emic* instead. An etic marker may be considered as a potential emic marker insofar as the interacting other might recognize the correlation of marker and state and use this information in further contacts. Necessary prerequisite is of course the possibility to notice the correlated state by means of other information. The distinction is important as the researcher who is interested in the communication process wants to learn about relevant information exchange between individuals. Therefore three different processes should be recognized in the analysis of interaction:

1. *Etic markers* that are correlated with a specific state and not perceived or recognized by the receiver;
2. *Emic markers* that are also correlated with a given state and are being recognized and processed as such;
3. *Attributions that are not based on actual markers.* These result from former interactions with this or other similar individuals or are innate. The human equivalent would be the attributions based on stereotypes.

Obviously emic vocal markers play an important role in communication, but which markers are clearly emic? If it is found, for example, that a given state is correlated with a certain frequency modulation in some calls of primates, this is then an emic marker for us, but it might as well be an etic marker for the interacting conspecific. Quite often etic markers are assessed, as these might be easy to measure (mean values of F0, timing, etc.). For the analysis of interaction it is important whether a marker is emic and what the relation is between markers and marked state. This is not an easy matter to demonstrate empirically (see contribution by Maurus et al., this Vol.). The distinctions presented here, derived from a semiotic framework might be useful for that purpose.

4 Analyzing the Process of Communication

We will now turn towards some points related to research strategies in studying vocal signalling in primates. In order to analyze the process of communication it is important to recognize several aspects of interaction:

1. Information transmission between individuals is indirect, it always requires a medium. Thus information is encoded by the sender and has to be decoded by the receiver. Consequently, the code as such is an important object of study.
2. Encoding and decoding are dependent on the individuals' habits, knowledge, experiences, etc.
3. Information transmission is not limited to one channel. Most often, several channels are used (sequentially and in parallel).
4. Communication is not context-free but occurs in a framework of cultural or social rules and information that is shared by the participating individuals.
5. A message may serve several functions as it has several meanings at different levels of interpretation (see above).
6. Information transmission may be disturbed at any point of the process (noise).

A modified Brunswikian lens model (K.R. Scherer 1986) may serve as a visualization of these notions (Fig. 1).

A trait (such as personality or status) or an organismic state (arousal or emotion) of the sender (C) is "externalized" ("encoded" or "expressed") via a series of distal indicator cues (D). In the case of vocalization, these cues (distal to the observer) consist of objectively measured acoustic parameters such as frequency, intensity, and timing. The assumption is that this externalization is mainly determined by push effects via

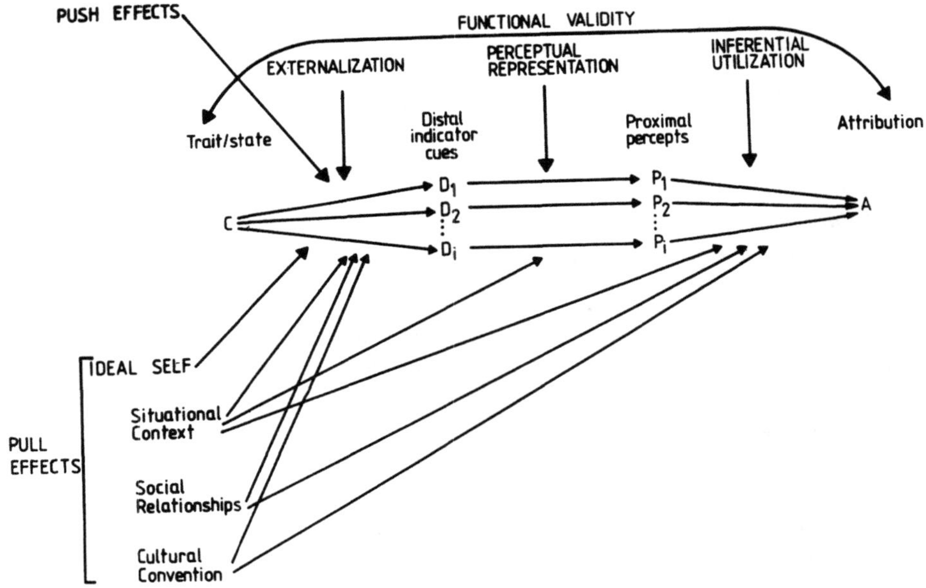

Fig. 1. Brunswikian lens model (revised)

physiological processes. However, even this component of signal production may be affected by various pull effects (in particular self-representation in the service of an "ideal" self). The distal cues are transmitted via acoustic waves and perceived by the sensorium of a receiver. There the acoustic parameters are represented as proximal percepts (C), in this case auditory impressions (e.g., subjectively perceived pitch, timbre, etc.). The transmission process is strongly affected by the situational context such as noise or other environmental factors. On the basis of the proximal percepts registered by the sensorium, the higher information processing systems of the receiver organism produce an attribution (A) concerning trait, state, and/or communication intention of the sender. This process can be called "inferential utilization of the proximal cues", even though the activity is not necessarily "cognitive" in the sense of neocortical activity. Clearly, this highly subjective inference/attribution process is strongly affected by schemas, expectations, and other framing structures based on situational context, existing social relationships, and cultural convention. The accuracy of the resulting attribution (in terms of the actual sender characteristics and intentions) are (after Brunswik) referred to as "functional validity".

The distinction between externalization (encoding) and perception (decoding) and the direction of the arrows in the model detract from the fact that these two aspects of signal use are highly interrelated. As has been shown by Leyhausen (1967) one can make a very good case for a shaping of signal structures by perception or impression requirements. As shown above, the interaction between push and pull effects is a particularly interesting issue as far as the expression of emotion is concerned.

Yet, the model seems useful as a paradigm for the study of vocal affect signalling since it encourages a more comprehensive approach, focussing on both the production of signals, the "expression", and the reception and use of signals, the "impression". Provided that the various parameters in the model can be objectively tested, correlation and path analysis methods can be used to test a number of hypotheses against the empirical relationships (see K.R. Scherer 1978, for examples). In particular, this approach is useful for examining the reason for communication failure: in examining the different sets of association coefficients in the model one can determine whether an inaccurate attribution of sender state/trait by the receiver is due to the lack of valid vocal cues (or etic markers), to transmission loss in the communication channel, to inaccurate perceptual representation of the acoustic parameters, or to faulty inference algorithms (inaccurate decoding).

5 Conclusions

The preceding discussion has shown that vocal affect signalling is an extraordinarily complex process and that the view that animal calls or human nonverbal vocal behavior are simple symptoms of internal affect or motive states is likely to be inaccurate. As Bühler's Organon model implies, signals are in part symptoms, but they are also symbols and appeals. As we have tried to show, the symptom aspect is likely to be determined by different factors (internal push) than the symbol and appeal aspects (external pull). The acoustic parameters of a vocalization most likely represent a combination of the respective determinants and thus represent both the underlying state of the sender as well as the nature of a conventional code and the specific context of a particular communication situation.

The claim for the existence of representational, symbolic signalling in animal calls (see above) can be well integrated into the model adopted here. Clearly, the vervet monkey alarm calls studied by Cheyney, Marler, and Seyfarth (see contribution by Marler and Mitani, this Vol.) are at the same time symptoms of fear, symbols of the type of predator, and appeals to conspecifics to take care. It seems possible that the symbol aspect is more pronounced in this example than in other types of animal calls given the strong adaptive advantage of clearly identifying the kind of predator to be beware of.

We will conclude this chapter with a speculation concerning the links between affect expression and symbolic representation. At first sight, one could assume that these are very different types of codes, an assumption that is widespread in psycholinguistics and semiotics. However, the component patterning model of affect expression outlined above suggests a way to conceptualize the emergence of symbolic representation from affect expression. If the basic assumption of the component patterning model is correct, i.e., the notion that the outcome of each stimulus evaluation check differentially changes the different organismic subsystems and thus produces communicative signals, then the complex signal expressing the affect state resulting from the evaluation should also reflect the evaluation process. In other words, the receiver of an affect signal should be able to reconstruct the evaluation sequence outcome on the basis of the effects of these outcomes on the stream of expressive behavior. Thus, a vocal affect signal should not only allow the receiver to infer the sender affect state in a relatively gross manner but also to reconstruct the evaluation process that led to the resulting affect state. Since the evaluation process, if it is accurate, reflects the nature of the objects or events that provoked the emotion-eliciting evaluation, the receiver should be able to reconstruct, at least in part, the essential dimensions of the eliciting objects or events. One could argue then that an affect expression is, in a rather abstract way, a "cognitive" representation of the affect-eliciting object or event since it reflects the evaluation process.

Let us take the vervet alarm calls as an example. According to the component patterning model, vocalization following the detection of a predator should reflect the following evaluation process: a sudden, new percept of another animal, intrinsically unpleasant because it belongs to the class of predators, hinders ongoing need/goal activity of feeding and endangers survival, consequences still seem controllable, fight and flight are still possible. If this evaluation outcome is inferred by a receiver on the basis of the vocal patterning of the alarm call, one could assume that it has obtained a fairly detailed representation of a particular class of objects relevant to this species. One could argue that this kind of analysis does not explain the differentiation between air and ground predators, given that the evaluation sequence may be the same for both at this level of abstraction. However, it would seem possible that such differences are coded in terms of the continuously varying scales of evaluation for each check. For example, air predators might appear more suddenly than ground predators and thus the differential result of the novelty check would lead to differential acoustic patterning. Other possibilities are differences in the urgency of adaptive response or the degree of control and power, requiring different response strategies. We should recall that the component patterning model assumes, following Darwin, that the change in expression is related to an adaptive response (in terms of a residue or an intention behavior). Consequently, we would expect different forms of expression if different predators require differential coping behavior.

The argument that affect expression signals may represent the evaluation sequence outcomes of the sender, does not negate the possibility of signal conventionalization and ritualization. On the contrary, one could assume that there is pressure to represent typical evaluation outcomes, and thereby the abstract dimensions of the objects or events producing the affect, more concisely and less affected by individual variation which may lessen the ability of the receiver to infer this information.

These concluding remarks have been very speculative and are offered in the spirit of generating further discussion about the determinants of vocal affect expression and its relationship to representational communication. We do hope that the three models emphasized in this chapter, Bühler's notion of multifunctional signals, the modified Brunswikian lens model relating expression to impression, and the component process model attempting to predict expressive patterning, may contribute to encourage more comprehensive, interdisciplinary research in this area.

References

Brown C (1982) Auditory localization and primate vocal behavior. In: Snowdon CT, Brown CH, Petersen MR (eds) Primate communication. Cambridge Univ Press, Cambridge, pp 144–164

Brown P, Levinson S (1979) Social structure, groups and interaction. In: Scherer KR, Giles H (eds) Social markers in speech. Cambridge Univ Press, Cambridge

Bühler K (1933) Ausdruckstheorie. Fischer, Jena

Caryl PG (1979) Communication by antagonistic displays: what can games theory contribute to ethology? Behaviour 86:136–169
Darwin C (1872) The expression of emotion in man and animals. Murray, London (Reprinted 1965, Univ Chicago Press, Chicago)
Dawkins R, Krebs JR (1978) Animal signals: information or manipulation? In: Krebs JR, Davis NB (eds) Behavioral ecology. Blackwell, Oxford
Eisenberg JF (1974) The function and motivational basis of hystericomorph vocalizations. Symp Zool Soc London 34:211–247
Eisenberg JF (1976) Communication mechanisms and social integration in the black spider monkey *(Ateles fusciceps robustus)* and related species. Smithsonian Contrib Zool 213
Ekman P, Friesen WV (1978) The facial action coding system: a technique for the measurement of facial movement. Consulting Psychologists, Palo Alto, Calif
Feldstein S, Welkowitz J (1978) A chronography of conversation: in defense of an objective approach. In: Siegman AW, Feldstein S (eds) Nonverbal behaviour communication. Erlbaum, New York
Giles H, Smith PM (1979) Accomodation theory: optimal levels of convergence. In: Giles H, St. Clair R (eds) The social psychology of language. Cambridge Univ Press, Cambridge
Giles H, Scherer KR, Taylor DM (1979) Speech markers in social interaction. In: Scherer KR, Giles H (eds) Social markers in speech. Cambridge Univ Press, Cambridge, pp 343–381
Goffman E (1959) The presentation of self in everyday life. Doubleday, Garden City
Goffman E (1969) Strategic interaction. Univ Philadelphia Press, Philadelphia
Goffman E (1979) Response cries. In: v. Cranach M, Foppa K, Lepenies W, Ploog D (eds) Human ethology: claims and limits of a new discipline. Cambridge Univ Press, Cambridge, pp 203–240
Gould E (1983) Mechanisms of mammalian auditory communication. In: Eisenberg JR, Kleiman D (eds) Advances in the study of mammalian behavior. Am Soc Mammal, New York, pp 285–341
Green S (1975) Variation of vocal pattern with social situation in the Japanese monkey *(Macaca fuscate)*: a field study. In: Rosenblum (ed) Primate behavior: developments in field and laboratory research, vol 4. Academic Press, London New York, pp 2–102
Hinde RA (1981) Animal signals. ethological and games-theory approaches are not incompatible. Anim Behav 29:535–542
Hooff J van (1972) A comparative approach to the phylogeny of laughter and smiling. In: Hine RA (ed) Nonverbal communication. Cambridge Univ Press, Cambridge, pp 209–241
Jürgens U (1979) Vocalization as an emotional indicator: a neuroethological study in the squirrel monkey. Behaviour 69:88–117
Jürgens U (1982) A neuroethological approach to the classification of vocalization in the squirrel monkey. In: Snowdon CT, Brown CH, Petersen RM (eds) Primate communication. Cambridge Univ Press, Cambridge, pp 50–62
Jürgens U, Ploog D (1976) Zur Evolution der Stimme. Arch Psychiat Nervenkrankh 222:117–137
Laver J (1980) The phonetic description of voice quality. Cambridge Univ Press, Cambridge
Laver J, Trudgill P (1979) Phonetic and linguistic markers in speech. In: Scherer KR, Giles H (eds) Social markers in speech. Cambridge Univ Press, Cambridge
Leyhausen P (1967) Biologie von Ausdruck und Eindruck (Teil 1). Psychol Forsch 31:113–176
Marler P (1965) Communication in monkey and apes. In: DeVore I (ed) Primate behavior. Holt, New York, pp 544–584
Marler P (1984) Animal communication: affect or cognition? In: Scherer KR, Ekman P (eds) Approaches to emotion. Erlbaum, New York, pp 345–368
Marler P, Tenaza R (1977) Signalling behavior of apes with special reference to vocalization. In: Sebeok (ed) How animals communicate. Indiana Univ Press, Bloomington, pp 965–1033
Morris CW (1946) Signs, language, and behavior. Prentice Hall, New York
Morton ES (1977) On the occurrence and significance of motivation-structural rules in some bird and mammal sounds. Am Nat 111:855–869
Ohala JJ (1980) The acoustic origin of the smile. J Acoust Soc Am 72:66
Peirce CS (1931–1935) Collected papers of Charles Sanders Peirce. In: Hartshorne C, Weiss P (eds) Harvard Univ Press, Cambridge, Mass

Ploog D (1974) Die Sprache der Affen und ihre Bedeutung für die Verständigungsweisen des Menschen. Kindler, München

Redican WK (1982) Facial displays of emotion by monkeys and apes: an evolutionary perspective on human facial displays. In: Ekman P (ed) Emotion in the human face, 2nd edn. Cambridge Univ Press, Cambridge, pp 212–280

Rowell TE, Hinde RA (1962) Vocal communication by the rhesus monkey. Proc Zool Soc London 2:279–294

Scherer KR (1977) Affektlaute und vokale Embleme. In: Posner R, Reinicke HP (eds) Zeichenprozesse – Semiotische Forschung in den Einzelwissenschaften. Athenaion, Frankfurt, pp 199–214

Scherer KR (1978) Personality inference from voice quality: the loud voice of extraversion. Eur J Soc Psychol 8:467–487

Scherer KR (1980) The functions of nonverbal signs in conversation. In: St Clair R, Giles H (eds) The social and psychological contexts of language. Erlbaum, New York

Scherer KR (1982) Methods of research in vocal communication: paradigms and parameters. In: Scherer KR, Ekman P (eds) Handbook of methods in nonverbal behavior research. Cambridge Univ Press, Cambridge, pp 136–198

Scherer KR (1985) Vocal affect signalling: a comparative approach. In: Rosenblatt JS, Beer C, Busnel MC, Slater PJB (eds) Advances in the study of behavior, vol 15. Academic Press, London New York, pp 189–244

Scherer KR (1986) Vocal affect expression: a review and a model for future research. Psychol Bull 99:143–165

Scherer KR, Giles H (1979) Social markers in speech. Cambridge Univ Press, Cambridge

Scherer U, Helfrich H, Scherer KR (1980) Internal push or external pull? Determinants of paralinguistic behavior. In: Giles H, Robinson P, Smith P (eds) Language: social psychological perspectives. Pergamon, Oxford New York, pp 279–282

Schlenker B (1980) Impression management: the self-concept, social identity, and interpersonal relations. Brooks/Cole, Monterey

Seyfarth RM, Cheney DL (1982) How monkeys see the world: a review of recent research on East African vervet monkeys. In: Snowdon CT, Brown CH, Petersen MR (eds) Primate communication. Cambridge Univ Press, Cambridge, pp 239–252

Tembrock G (1971) Biokommunikation: Informationsübertragung im biologischen Bereich. Akademie, Berlin

Tembrock G (1975) Die Erforschung des tierlichen Stimmausdruckes (Bioakustik). In: Trojan F (ed) Biophonetik. Bibliogr Inst, Mannheim

Williams CE, Stevens KN (1972) Emotions and speech: Some acoustical correlates. J Acoust Soc Am 52:1238–1250

Zahavi A (1982) The patterns of vocal signals and the information they convey. Behaviour 80:1–8

Neurobiology and Pathology of Subhuman Vocal Communication and Human Speech

DETLEV PLOOG [1]

The ability to communicate acoustically is shared by man with many vertebrates. The ability to speak, however, is shared with no other living creature, it is species-specific to *Homo sapiens*. There exists a close functional relationship, however, between voice and speech; they operate as a tandem in vocal communication. Can we learn something about the development of this relationship and the interplay between phonation and speech by studying our closest relatives, the subhuman primates?

This contribution is focussed on a comparison of the major peripheral and central nervous structures and functions which man and monkey have in common and of those in which they differ. A description of the fundamental differences will be based on results obtained by brain lesions and brain stimulation experiments in monkeys and man and on clinical cases of speech pathology (Ploog 1985).

Figure 1 represents a variation of the figure Jürgens (see p. 165) used in explaining the hierarchical central control of monkey calls involving organizational levels, structures, and functions with which this chapter is also concerned. Regarding functional level IV, Fig. 1 has been complemented insofar as it shows a lateral view of a primate brain with the cortical representation of the larynx, pharynx, tongue, and lips (LR).

The dotted area above the cingulate gyrus (GC) indicates the supplementary motor area (SMA), a detailed discussion of which will follow. Anatomically, and with one exception, all structures shown in this figure are homologous in man and monkey, even though some of them differ in function. The one exception, as mentioned already by Jürgens, is the direct motor pathway from the laryngeal representation in the primary motor cortex to the laryngeal motoneurons in the medulla which does not exist in the monkey (Kuypers 1958; Jürgens 1976). This direct connection serves as the neuronal basis for the voluntary control of the vocal folds in man which is not possible in the monkey. However, the final pathway from the medullary motor nuclei to the phonatory muscles is homologous in man and monkey. The intrinsic laryngeal muscles are innervated by two motor nerves originating from the nucleus ambiguus. Transection experiments in the squirrel monkey revealed that the functional role played by these nerves in phonation is similar in man and monkey (Jürgens et al. 1978).

1 Max-Planck-Institut für Psychiatrie, Kraepelinstr. 2, 8000 München 40, FRG

Todt/Goedeking/Symmes (Eds.)
Primate Vocal Communication

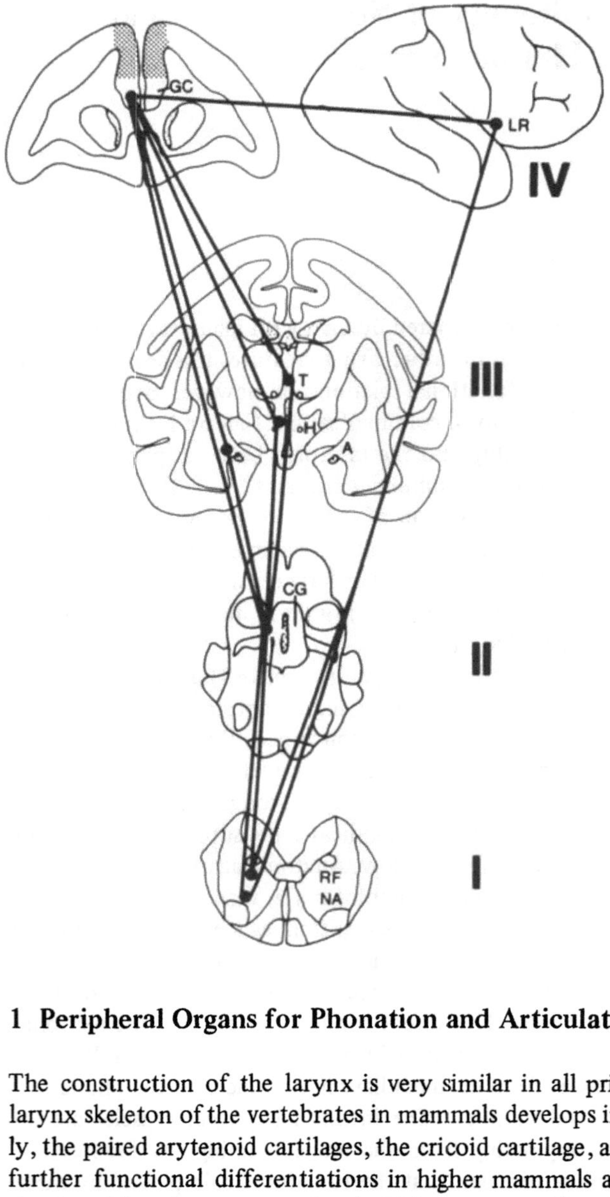

Fig. 1. Hierarchical control of voice and speech. *LR* Representation of larynx and articulatory organs in the neocortex; *stippled* supplementary motor area; *GC* gyrus cinguli (anterior limbic cortex); *T* thalamus; *H* hypothalamus; *A* amygdala; *CG* central grey; *RF* formatio reticularis; *NA* Nucleus ambiguus

1 Peripheral Organs for Phonation and Articulation

The construction of the larynx is very similar in all primates. The tripartite primary larynx skeleton of the vertebrates in mammals develops into a quadripartite one, namely, the paired arytenoid cartilages, the cricoid cartilage, and the thyroid cartilage. After further functional differentiations in higher mammals a functionally essential muscle developed in primates, namely, the M. interarytenoideus located between the arytenoid cartilages. The M. vocalis, in subhuman primates located at the base of the vocal cords only, in man reaches into the medial part of the vocal cords, making a finer adjustment of the voice possible (Bless and Abbs 1983). Again, in contrast to subhuman primates, vocal adjustment in man is under volitional cortical control. Without this control, man would not be able to sing, for instance.

The improvement of the phonation apparatus of vertebrates in the course of evolution seems to indicate that acoustic communication offered a selective advantage, i.e.,

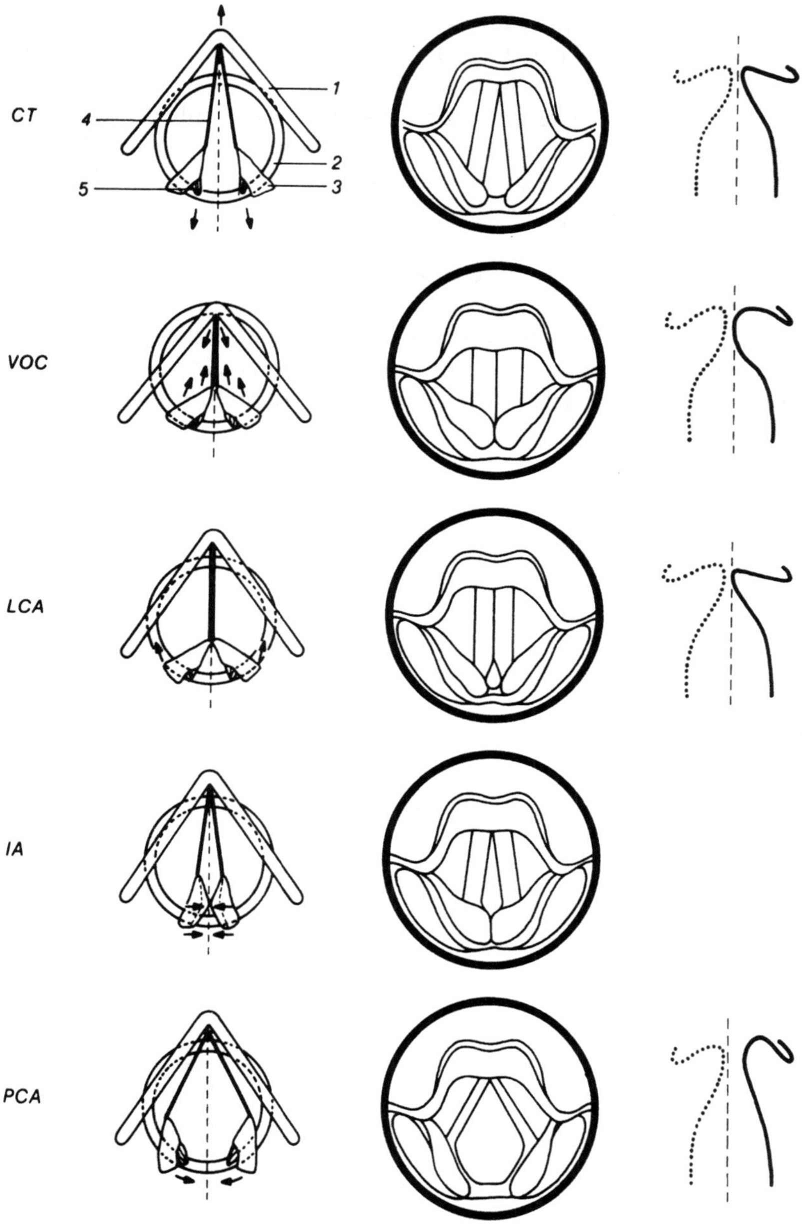

Fig. 2. Schematic presentation of the function of the laryngeal muscles. The *left column* shows the location of the cartilages and the edge of the vocal folds when the laryngeal muscles are activated individually. The *arrows* indicate direction of the force exerted. *1* Tyroid cartilage; *2* cricoid cartilage; *3* arytenoid cartilage; *4* vocal ligament; *5* posterior cricoarytenoid ligament. The *middle column* shows views from above. The *right column* presents contours of the frontal section at the middle of the membranous portion of the vocal fold. The *dotted line* shows a control where no muscle is activated (Hirano 1981)

that a rich repertoire supported speciation. The highly differentiated performance of the CNS with regard to the glottal movements is depicted in Fig. 2 by a schematic drawing of the functions of the five interior laryngeal muscles.

Moving the glottis apart in this interaction is the responsibility of only one abductor, the M. cricoarytenoideus posterior; all the other muscles serve to extend, shorten, tense, and close the vocal cords. In other words, what is taking place is a complex synergism of several muscles acting in close coordination with the respiratory and the external laryngeal muscles.

Figure 3 makes it evident that the differences between man and monkey are most obvious in the pharynx from the epiglottis to the cavity of the mouth. This region, the supralaryngeal tract, functionally of decisive importance to human articulation, is considerably larger in subhuman primates and particularly so in man. In the chimpanzee, the angle between pharynx and epiglottis is shaped like a flat arch, in a human adult,

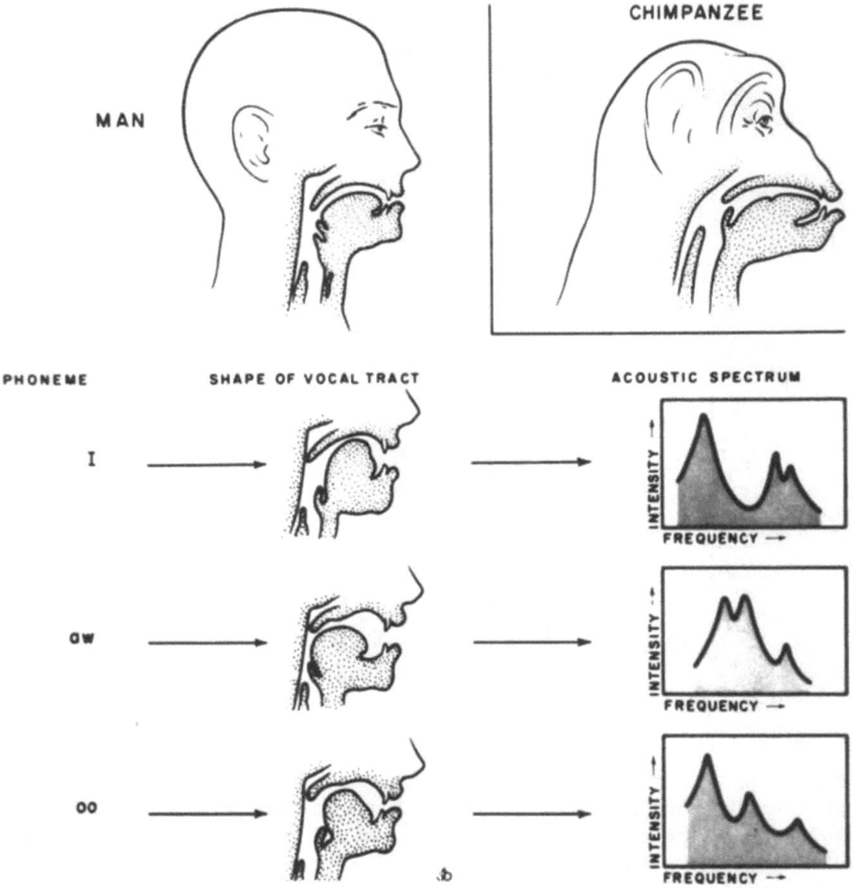

Fig. 3. The phonation apparatus of man and chimpanzee. The man-inherent configuration of the vocal tract enables him to produce sounds that the chimpanzee cannot produce, e.g. the three vowels presented in their articulatory positions and their intensity-frequency spectrum (Wilson 1975)

however, it is almost rectangular so that epiglottis and soft palate lie far apart. The lower part of the figure demonstrates why it is possible only for man to form such clear vowels as [i], [a] and [u]. Ontogenetically, this configurational state is reached at an age of 6 months and is completed during the years of speech development. The vocal tract of children with a marked Down syndrom has hardly changed since their early infancy which makes them unable to speak (Benda 1969; Lieberman 1973). Coupled with the development of the human vocal tract is a lowering of the tongue base. This, in addition to several morphological factors, is responsible for the tongue's extraordinary volubility.

In order to appreciate the complex motor activities in the vocal tract, let us consider only one aspect of sound production, the positions of the active articulatory organs, which are the lips, the jaw, the tip and the back of the tongue, at the time of production in relation to points at the palate between upper lip and uvula (Fig. 4). During speech these positions are shifted rapidly and, at the same time, the degree of openness of the vocal tract is varied. By a concerted action of these articulatory gestures with laryngeal movements the airstream produced by the lungs is transformed into a sound wave of high temporal and spectral complexity. In perception this continuous signal is resolved into a sequence of discrete entities, the so-called phonemes, each of which can be related to a particular configuration of the speech organs. The phonemes serve as a basis for a diversified phonetic classification system which can be arranged in a limited number of categories according to the type of articulation. Sounds produced with the mouth passage occluded and velum lifted, for instance, are categorized as plosives, those produced with the mouth passage occluded and velum lowered are nasals. By varying the degree of opening at the articulation points, the active organs can produce fricatives, vibrants, semivowels, or vowels. The interplay of vocal cord activity, type

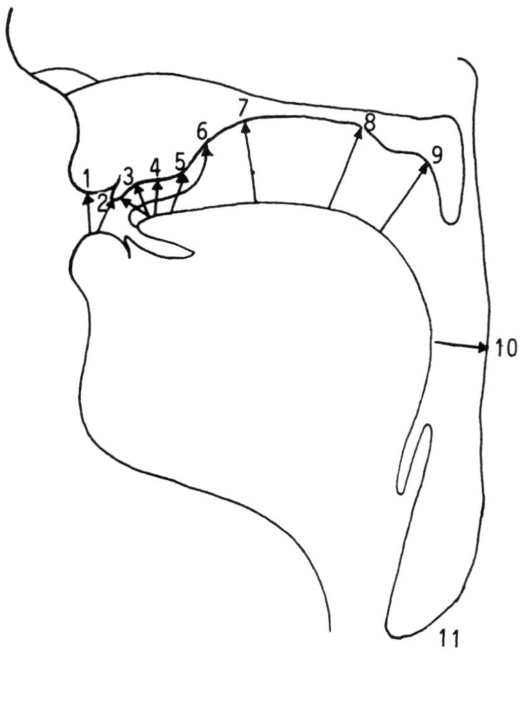

Fig. 4. Schematic drawing of articulation sites: *1* bilabial; *2* labiodental; *3* dental and interdental; *4* alveolar; *5* postalveolar, palatoalveolar; *6* retroflex; *7* palatal; *8* velar; *9* uvular; *10* pharyngeal; *11* glottal

and position of articulation makes a threefold classification possible. [P], for example, is a voiceless, bilabial plosive, [v] a voiced, labiodental fricative, [n] a voiced alveolar nasal sound. However, there is not only this relatively static approach which considers all sound-producing parameters involved in the production of a certain phoneme. When speaking, a great number of phonemes are produced so rapidly that sounds of different categories are combined, are even overlapping each other (coarticulation). Depending on the phonetic neighborhood (initial, center, or end position of a phoneme in a word, in a sentence, between sentences), the sound-producing parameters of the phonemes, e.g. the articulation sites, experience smooth changes in accordance with combinatorial rules. The effect of these highly complex synergetic activities enables the recipient of the vocal signal to decode and understand it regardless of its "make" (Lindblom 1983).

Speed and temporal arrangement are of great importance in all motor performances involving precision. And this is particularly true for speech, the motor performance requiring more precision than any other. Normally, 220 syllables are spoken per minute, but 500 have also been registered, and an average of about 14 phonemes are produced per second (Lenneberg 1967). Although we are not certain as to the exact number of muscles participating in the production of speech, we do know that an interaction of the muscles of the abdomen, thorax, throat, face, larynx, pharynx, jaw, tongue, and lips is involved. How the CNS generates, coordinates, and controls these synergetic motoric configurations is a problem still unsolved.

2 Central Nervous Organization of Phonation

The squirrel monkey *(Saimiri sciureus)* has been studied more thoroughly than any other primate with regard to vocal repertoire and brain structures involved (Jürgens 1979a,b; Ploog 1981; Jürgens and Ploog 1981).

Figure 5 shows four schematized sagittal views of the monkey's brain. Below each diagram are the sonagrams of various vocalizations elicited from the brain structures marked black by electrical brain stimulation (Jürgens and Ploog 1970; Jürgens 1979a,b). The calls are identical to those recorded under "natural" conditions. All natural calls are genetically preprogrammed in their acoustic structure and typical for the species like fixed action patterns. Learning plays a role only in connection with the use of vocal signals in the social context (Winter et al. 1973; Maurus et al. 1985). Jürgens stated that the electrical brain stimulation of structures which are part of the limbic system, or are directly connected with it, yields secondary vocal reactions expressing the specific motivational state.

In man, too, these limbic structures can be involved in nonverbal, emotional utterances, especially if means of verbal expression are blocked, e.g. in cases of motor aphasia (Robinson 1976), under the strong influence of emotions, or functionally uncoupled as in dreams (Hobson and McCarley 1977) where the dreamer can sigh, moan, scream, and laugh. The nonverbal utterances of small children not able to speak yet, utterances which can certainly have intentional characteristics, and the nonverbal emotional utterances of adults are based on the same neuronal mechanisms as the vocalizations of subhuman primates.

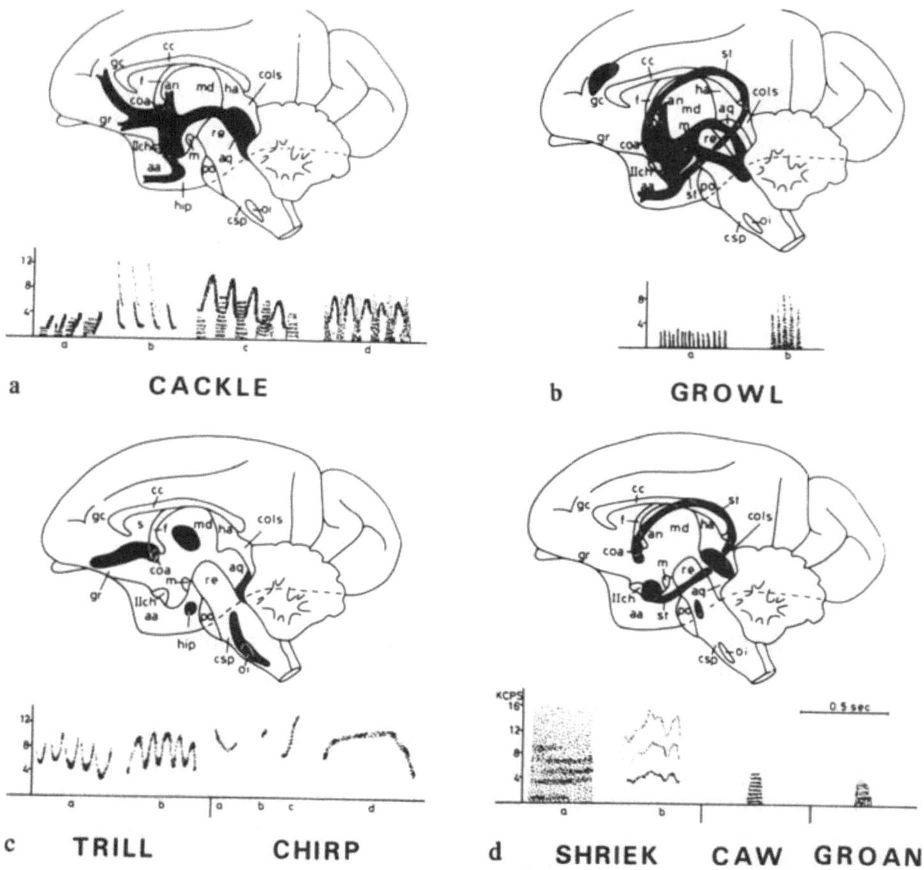

Fig. 5a–d. Cerebral structures of the squirrel monkey (*black*) yielding species-specific vocalization when electrically stimulated. *Below:* sagittal view, sonagrams (frequency, time) of respective call types

3 The Central Gray and the Traumatic Midbrain Syndrome

In the monkey's subcortical system yielding vocalizations when electrically stimulated there are two regions which are of particular functional significance to both monkey and man: the central gray in the midbrain and the anterior cingulate gyrus (see Fig. 1, II and IV). In Fig. 6 the sagittal view of a human brain is shown where the location and extent of each of these structures are marked. As Jürgens pointed out, destruction of the central gray causes mutism in the squirrel monkey (Jürgens and Pratt 1979). Also in man mutism can be caused by a closed head trauma involving damage of the periaqueductal gray (or its projections to the nucleus ambiguus). This so-called traumatic midbrain syndrome leads to long-lasting serious defects of phonation and articulation (Vogel and von Cramon 1982, 1983; Morasch and von Cramon 1984). During recovery three phonetic phases can be distinguished: the aphonic phase in which volitional adduction of the vocal cords, so as to produce sounds, is impossible for the patient, even though the reflectory vocal cord movements are not affected. Weeks, even months

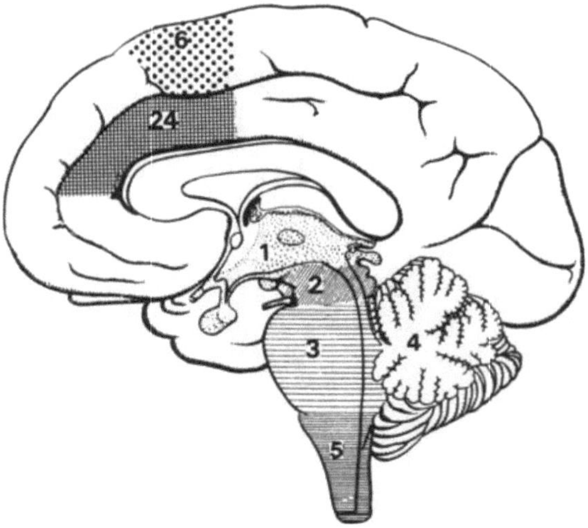

Fig. 6. Sagittal view of human brain; location and proportion of areas mentioned in text: *1* diencephalon; *2* mesencephalon; *3* pons; *4* cerebellum; *5* medulla; *6* area 6, supplementary motor cortex; *24* area 24, gyrus cinguli anterior

after this phase of laryngeal akinesia the patient begins to whisper volitionally articulated utterances. Soon thereafter the third phase begins in which he produces, sometimes under great strain, voiced sounds, i.e., periodic vocal cord vibrations.

Regarding the articulatory movements, the recovery process also develops characteristically. The phase of mutism is marked by anarthria. No articulation movements of any phonological significance can be performed. A phase of rudimentary articulation activities follows which at first involves primarily the area of the lips and the tip of the tongue. During the third phase, the vocal repertoire increases; labial sounds are produced first, the velar consonants last. The remaining impairments of voice and articulation are categorized according to the type of disorder, such as rigid, spastic, or atactic dysarthrophonia. Each type is characterized by a number of phonological features differently effecting the communication abilities of the patient (Ziegler and von Cramon 1988). Familiarity with these disorder patterns is important for logopedic rehabilitation. Which neuronal processes eventually lead to restoration, if only in part, of the patient's vocal communication is still unknown.

4 The Cingulate Gyrus – Its Involvement in Human Phonation

We know that after destruction of the cingulate gyrus, the monkey is no longer able to master a vocal operant conditioning task, that is, to vocalize in order to get a food reward, whereas communication in the communal situation remains unimpaired (Sutton et al. 1974; Kirzinger and Jürgens 1982). It is concluded that the anterior limbic mesocortex is responsible for the voluntary initiation of the voice. How can this situation be compared to man? Here is an example (Jürgens and von Cramon 1982; Fig. 7).

The 41-year-old patient suffered a cerebral infarction caused by occlusion of the ascending branches of both anterior cerebral arteries which affected the anterior cingu-

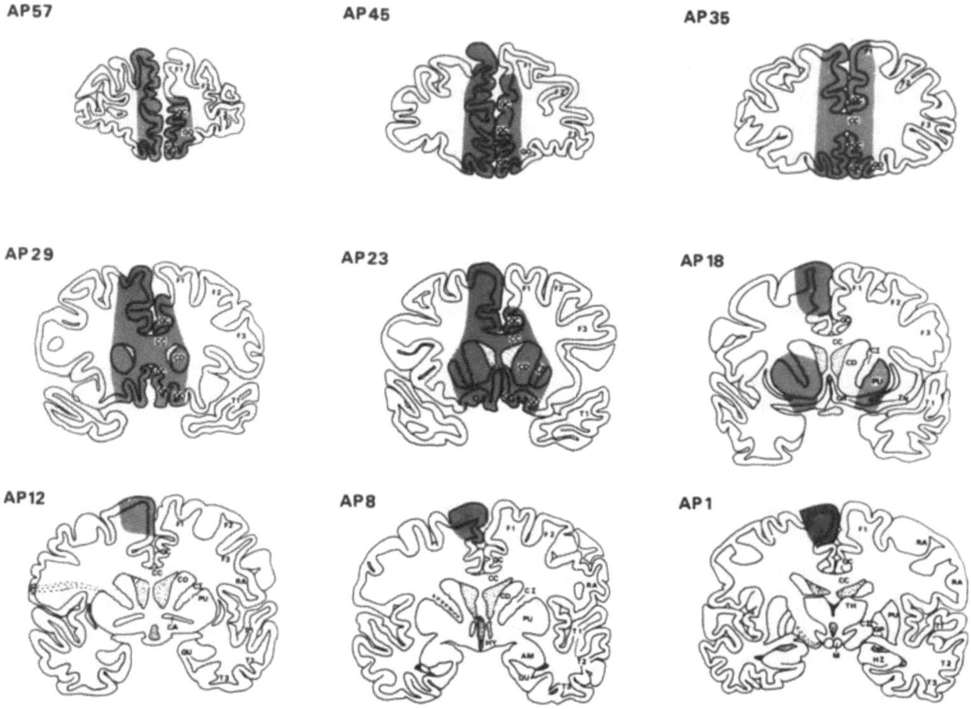

Fig. 7. Schematic view of human brain at nine levels from front (*AP 57*) to rear (*AP 1*). *Grey zones:* cerebral infarction sites, *AP 12, AP 8, AP 1*. *Dashed lines:* motor pathway from cortex to larynx and articulatory organs (Jürgens and von Cramon 1982)

late cortex bilaterally, the left supplementary motor area, the medial orbital cortex bilaterally, and the most rostral part of the striatum bilaterally. During the first 6 weeks the patient remained in a state of akinetic mutism; only his eyes were open. Occasionally, he responded to painful stimulus with a moan, moved his head or arms, or tried to smile. After 10 weeks, he repeated long sentences in whispers but articulating them clearly and making no mistakes. Spontaneous utterances, however, remained reduced to a few monosyllabic words. After a year his speech was still restricted and monotonous, even though he made no mistakes. In an intonation test performed after 5 years, he was not able to attach the appropriate emotion to certain short sentences, such as "shut up", "super", or "oh". The sonagrams revealed that his ability to speak with emotion was greatly reduced and that he was not able to correct this deficiency.

Responsible for this lack of prosody and the inability to control intonation is probably the bilateral lesion of the anterior cingulate cortex. It may well be possible that control of emotional expressions is exercised by the many anatomic projections going from there to other limbic structures (Fig. 1). Homologous to the deficiency in our patient is the inability of a lesioned monkey to perform vocally in a conditional test. However, while the monkey was still able to communicate vocally with his conspecifics, the lesioned man had great difficulties using his voice for speaking, although he was able to groan (phonation) and to whisper (articulation).

However, information gained only from studying the case of this patient does not suffice to form a definite opinion as to the controlling function the supplementary motor cortex has in phonation. Based on the clinical literature dealing with this subject, it is difficult to say whether and how it affects speech. A lesion of the supplementary motor cortex can have similar effects as in our example. If this section is removed on the side of the speech hemisphere, temporary aphonia will result (Penfield and Roberts 1959).

5 The Cortical Face Area – Its Involvement in Human Phonation

The next step in our comparison of subhuman vocal communication with human speech takes us to the lateral neocortex (Fig. 1, IV), the structure indispensable for speaking. A second example is to demonstrate that, in man, part of the cortex where larynx, pharynx, tongue, and mouth are represented is necessary for controlling the voice, whereas the monkey, as stated by Jürgens (this Vol., p.), does not need these structures for phonation (Fig. 8).

The 52-year-old, right-handed patient suffered an embolic cerebral infarction in the stem of the left middle cerebral artery. He had a hemiplegia on the right side, a right-sided lower facial paresis, and tongue deviation to the right. After returning to consciousness, he was fully oriented, understood what was said to him but could not utter a sound except that he coughed when the base of his tongue was touched. A laryngoscopical examination revealed that during respiration as well as phonatory attempts both vocal folds remained motionless. This state of complete mutism and inability to phonate lasted for about 11 weeks. Then, similar to the patient in our first example, the patient began to whisper words, and after 2 weeks phonation was almost completely restored. The voice, however, was monotonous and the quality of singing mediocre. In contrast to the situation in the first example, the patient in this case had considerable difficulties with articulation and the ability to make oral movements.

The brain lesion of the patient (Fig. 8a) was then experimentally modeled in the monkey (Fig. 8b) (Jürgens et al. 1982).

The lesion of the patient invaded Broca's area, the inferior pre- and postcentral cortex, rolandic operculum, inferior parietal cortex, insula, claustrum, and large parts of the putamen and white matter underlying the inferior frontoparietal and insular cortex.

Regarding primate vocalization, there is no evidence that one hemisphere exercises a greater influence on phonation than the other. Vocalizations are elicitable from areas either in the left or the right hemisphere. So the lesion in the monkey was carried out bilaterally. It invaded the homologue of Broca's area, the inferior pre- and postcentral cortex, rolandic operculum, inferior parietal cortex, and large parts of the insula, claustrum, putamen, and white matter underlying the inferior frontoparietal and insular cortex. In contrast to our patient, there was also a bilateral encroachment on the superior temporal gyrus and the internal capsule of one side. These findings, however, are of no relevance for our comparative purposes.

Although the monkey's tongue, lips, and masticatory muscles were completely paralyzed after the operation so that he could neither bite, chew, nor lick, he could swallow and his phonation remained intact. The spectrographically registered vocaliza-

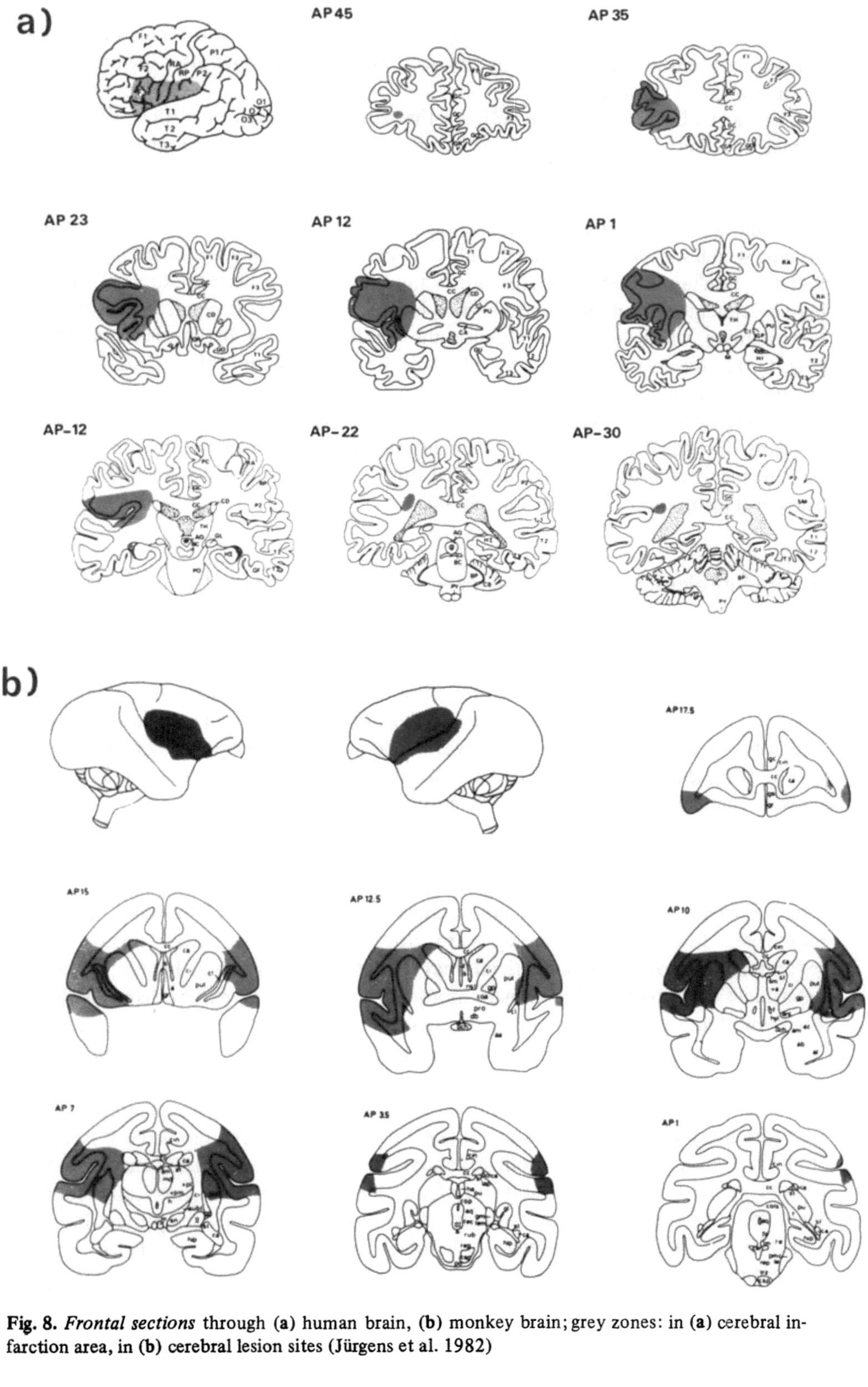

Fig. 8. *Frontal sections* through (**a**) human brain, (**b**) monkey brain; grey zones: in (**a**) cerebral infarction area, in (**b**) cerebral lesion sites (Jürgens et al. 1982)

tions included all call types described above. Consequently, the vocal folds were functioning, i.e., central nervous control was not impaired. Four days after the operation, the animal could chew again.

The most obvious difference between the patient's condition and that of the test animal was that despite the bilateral lesion which resulted in a complete paresis of tongue and mouth, the monkey kept his ability to vocalize, while the patient, despite the unilateral lesion, remained aphonic for 11 weeks. In other words, he could neither utter emotional nonverbal sounds, such as laughing, crying, moaning, nor articulated sounds, which means that his vocal folds were completely paralyzed. Based on the fact that the monkey did not have this defect and on additional neuroanatomical data, it may be concluded that the human cortex, in which the larynx is represented, contains motoneurons which directly influence the motoneurons in the nucleus ambiguus (Fig. 1). The observation that the vocal folds of the patient, despite the unilateral lesion, remained motionless bilaterally is in accord with earlier reports of similar cases where initial mutism was always connected with a left-sided, never a right-sided cerebral lesion. It seems that the left hemisphere, in man, exercises a controlling function over the right hemisphere with regard to vocal fold functions, a control which gradually diminishes with restoration of phonation abilities.

6 The Supplementary Motor Area – Its Involvement in Human Speech

What remains to be discussed in correspondence with Figs. 1 and 6 is the function of the supplementary motor area (SMA) in monkey and man. Jürgens (1984), in a comprehensive anatomical study of the efferent and afferent connections of this area, was able to show that the SMA projections on the cortical level as well as onto subcortical structures of the vertical brain axis down to the spinal level is closer related to motor-coordinating than to tactile, visual, or auditory sensory analyzing structures. There are massive connections with the primary motor cortex, premotor areas 6 and 8, striatum, subthalamus, ventrolateral thalamus, nucleus ruber, and spinal cord, all of which have been related with one or the other aspect of motor control. The only more direct sensory input to SMA comes from proprioception. A comparison of the efferent and afferent SMA projections show an almost complete reciprocity in prefrontal area 9, premotor areas 6, 8, and 44 (homologous to Broca's area!), orbital cortex, cingulate cortex, primary motor area 4, and parietal areas 2, 5, and 7. According to Jürgens, the SMA determines the activity of the primary motor cortex to a greater extent than vice versa. Anatomical and physiological evidence suggests that the SMA not only participates in the initiation of global motor programs but that its influence also reaches into the motor subroutines.

As far as the voice is concerned, there are several experiments which strongly support this interpretation of SMA functioning. For example, squirrel monkeys, if isolated from their group, utter series of long-distance contact calls, the so-called isolation peeps (Winter et al. 1966). Bilateral ablation of the anterior part of SMA drastically reduces the number of this type but not any one of the other call types (Kirzinger and Jürgens 1982). In this situation SMA seems to be involved in the initiation of a behavior pattern not directly triggered by external stimuli. Furthermore, Sutton et al. (1981) trained monkeys to press a lever during presentation of one light and to utter a vocalization during presentation of another light. Correct responses in both cases were food-rewarded. After ablation of SMA, there was a sustained increase in latency of the vocal operant response, whereas the lever-pressing response did not show any deficit. It seems that it is the type of motor response, rather than that of the eliciting stimulus, which determines the deficit. As vocal operant responses are much more difficult to learn for monkeys than lever-pressing responses, the deficits after SMA lesions in this case correlate with what Jürgens (1984) calls "volitional effort" necessary to produce a certain behavior pattern.

In conclusion, it may be stated that the SMA, and probably only its anterior part, is the only neocortical area involved in vocal behavior of nonhuman primates. In man, this area plays an important role in speech production. Its destruction in the dominant hemisphere (and with a minor effect also in the nondominant hemisphere) causes transient mutism and longer lasting dysarthrophonia (see case report 1, p. 202f.) (Penfield and Roberts 1959). The most direct evidence for the active participation of SMA can be demonstrated by regional blood flow measurements.

7 The Perisylvian Cortex – Its Involvement in Speech and Language

Shown in Fig. 9 is a left-sided intensive activation of the prefrontal/premotor region which corresponds to the supplementary motor cortex (area 6 in Fig. 6), as well as a second activation of the larynx-tongue area (LR in Fig. 1). To a limited degree, the right hemisphere participates in this activation. Measurements were taken while the healthy test persons, in so-called automatic speech tests, repeatedly counted from 1 to 20. A similar activation of area 6 could be observed when complex movements were executed, e.g., a sequence of isolated finger movements; this activation occurred even if the test person only imagined these movements instead of actually carrying them out (Lassen and Larsen 1980; Ingvar 1983).

In addition to area 6, other prefrontal/premotor areas (Brodmanns area 8 and 9) are activated in sequential motor tests, when articulating sequences of words, for instance. The prefrontal activation takes place even if the test person does not speak but thinks only of word groups, e.g. weekdays, numbers, or poems. This is called "inner" speaking (Fig. 10) which takes us to the field of ideation and its possible measurability.

Based on the data collected at the clinic of cerebral speech disorders an activation of the classical speech center of Broca (Fig. 8, F3) may be assumed. Which role this area, extending in front of the face area (LR), plays in speech has become a matter of dispute again. It may be conceivable that there is a stage in the child's development where the "raw material" of innate vocalization, e.g. a child's cooing and babbling, is assembled and formed into phonemes by Broca's area, released via the face area (LR) by vocal gestures learned by imitation and screened against motorically active neighboring cortex areas.

The conspicuous associative movements of the extremities, especially the hands, which accompany the articulatory movements of a child learning to speak may be indicative of a functional differentiation process taking place within and between the respective areas of the motor and premotor cortex.

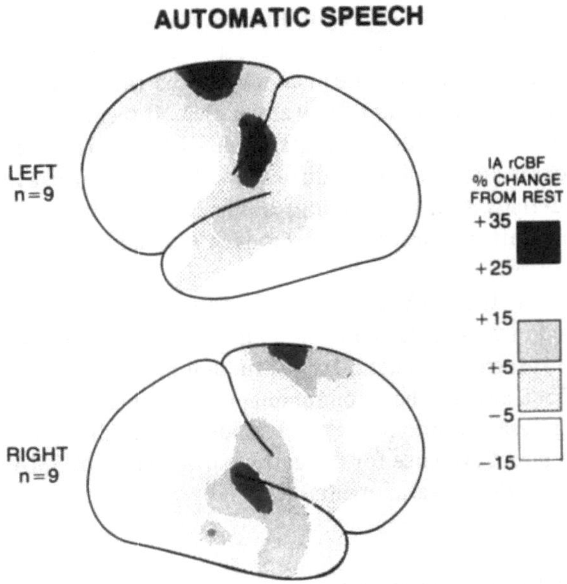

Fig. 9. Automatic speech. Superimposed diagrams of nine right-sided and nine left-sided IA rCBF studies in patients without neurological disturbance and with normal speech. The rCBF changes have been calculated in percent relative to the resting state. *Scale to the right* denotes magnitude of flow change. During automatic speech the subjects were asked to count repeatedly from 1–20. The subjects had their eyes closed. Note z-like flow change on the *left side* with a clear-cut flow peak in the premotor/prefrontal regions, another peak in the mouth/tongue/larynx area, and also an activation of the middle temporal region. On the *right side* a similar pattern was recorded, but the peaks were not as high, and less well defined, especially in the temporal region. Replotted after color TV display (Ingvar 1983)

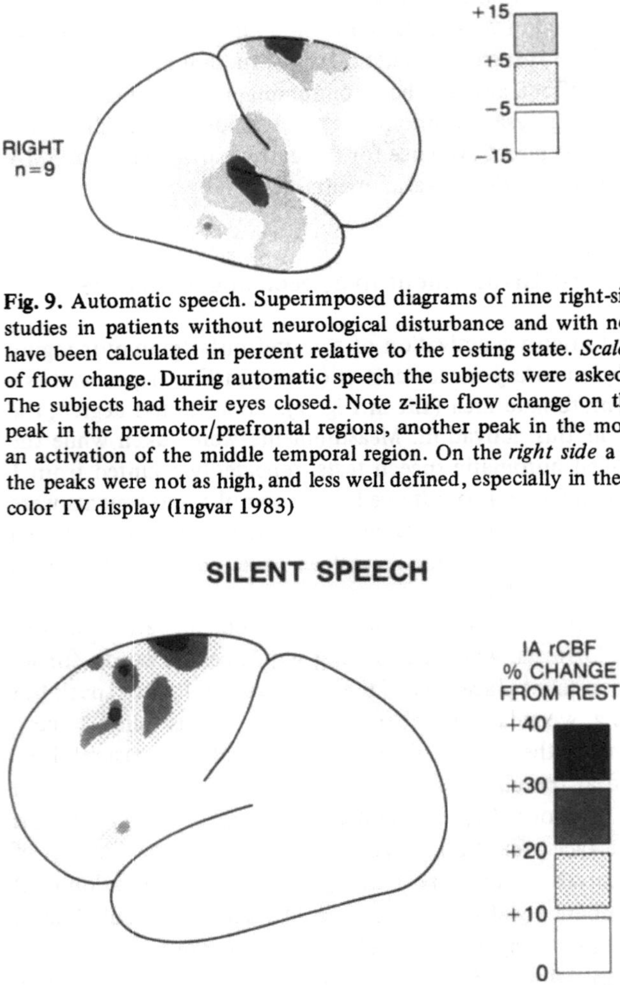

Fig. 10. Silent speech. IA rCBF measurement with a high resolution instrument with 254 detectors in the left hemisphere in a normal subject. Flow changes displayed in percent related to the resting value for the respective region in accordance with the *scale to the right*. During silent ("inner") speech the subjects were asked to count from 1–100 silently. Occasional questions from an observer established how far the subject had come. Note increase of flow in *upper prefrontal regions*. Replotted from color TV display (Larsen et al. 1978)

The final part of this contribution deals with results of electrical brain stimulation in man as far as speech and language functions are concerned. We know that it is not possible to elicit vocalizations by electrical stimulation of the monkey's neocortex. Perhaps the only exception is the chimpanzee. In experiments with this animal it was sometimes possible to evoke brief sounds (Leyton and Sherrington 1917; Hines 1940). Stimulating this area in the squirrel monkey will lead to vocal fold movements but not to vocalizations (Jürgens 1976a,b). In man, indifferent vocalizations may be evoked by stimulating the area of LR and SMA (Penfield and Roberts 1959). It appears that even an indistinctive excitation of the vocal system can induce a certain coordination between vocal folds and exspiration.

A modified form of electrical brain stimulation for the purpose of investigating language functions at the cortical and subcortical levels was introduced by Ojemann and co-workers (Ojemann and Mateer 1979). It is based on the neurophysiologically well-founded fact that electrical brain stimulation not only causes local excitation but also local inhibition. Especially this latter effect was used by Ojemann in identifying a common perisylvian cortex for motor and language functions in the language dominant hemisphere, including sites common to sequencing of motor movements and comprehension of phonemes (Fig. 11; Ojemann 1982).

The following language-related functions were studied: object naming, reading of simple sentences, short-term verbal memory, imitation of single and sequential orofacial movements, and phoneme identification. The latter involved the patient identifying aloud stop consonants embedded in a test phrase.

Stimulation effects such as these tasks are most often disruptive, ranging from failure to perform at all (of what was performed before stimulation) to production of wrong words or incorrect sequences. Most of the data come from the study of stimulation effects in the lateral cortex of the dominant hemisphere on naming. But also in the thalamus there are discretely localized language functions, with changes in naming, confined to medial central portions of the ventrolateral thalamus and anterior superior lateral pulvinar. Several types of naming errors are unique to thalamus: perseveration at sites in the midportion of medial ventrolateral thalamus and repeated use of the same wrong word anteriorly (Ojemann 1982).

Language and motor functions share common sites in the perisylvian cortex. These include sites in the posterior-inferior frontal lobe related to a motor output mechanism for all speech. They are distinguished from the face motor cortex by the absence of evoked movement. Rather unexpectedly, comprehension of phonemes is also often altered at these sites. Ojemann draws the remarkable and, for current speech theories, most important conclusion that "this common comprehension of speech sounds provides an appropriate anatomic substrate for the motor theory of speech perception" (Liberman et al. 1967). That theory suggests that speech sound decoding involves creation of a motor representation of the sound, the two function stimulation localizes to this common cortex.

This brings us back to the mechanisms of the voluntary control of the voice which seem to be a prerogative for the motor matching device of Liberman et al. (1967). This faculty has been achieved in humans only. Several preconditions and precursors of this faculty, some of which were depicted in this contribution, are present in the nonhuman primate.

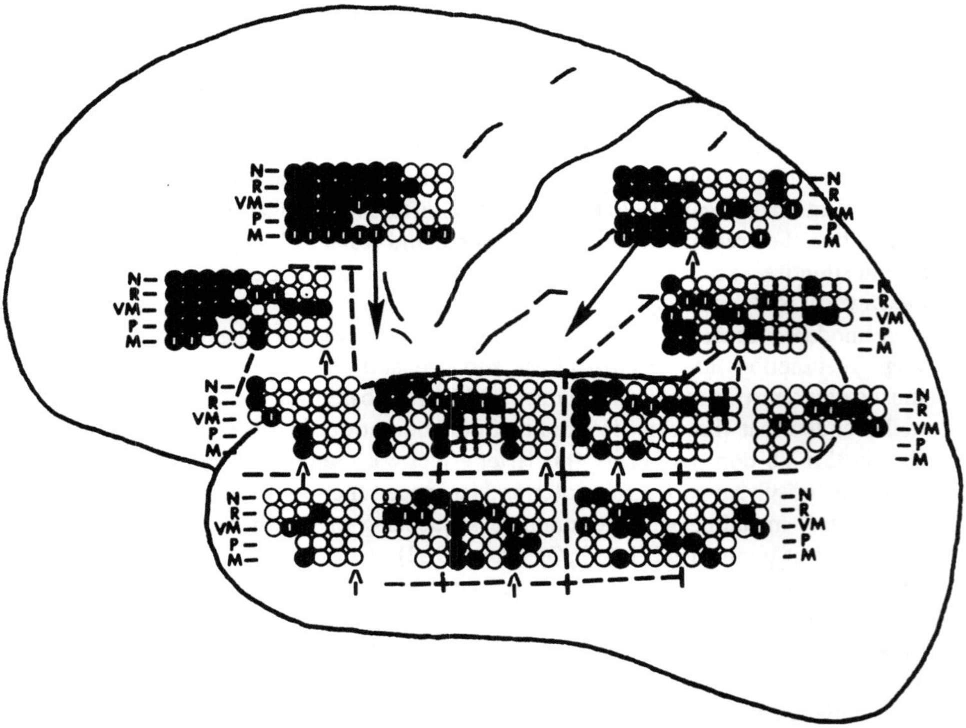

Fig. 11. Stimulation evoked changes in naming (*N*), reading (*R*), short-term verbal memory (*VM*), phoneme identification (*P*), and mimicry of single and sequential orofacial movements (*M*) in 11 zones of the lateral cortex in the left, language-dominant hemisphere of 14 patients. Stimulation sites in each patient assigned to a zone by the relation to the motor cortex and end of the Sylvian fissure. Boundaries of each zone indicated by *dashed lines*. Performance within each zone represented by a *matrix*. Each *column* represents the performance of a patient with one or more sites in that zone. Each *row* is the performance on one of the language-related functions tested. An *open circle* indicates that no significant errors were evoked in that zone for that patient on that function; a *filled circle* represents significant evoked errors for that patient in that zone on that function. *Absence of a circle* indicates that the function was no tested in that patient in that zone. *Vertical white bars in filled circles* indicate special types of errors: errors only in syntax on reading, errors only with stimulation during input or storage on verbal memory, and errors on mimicry of even single movements. The eight *smaller arrows* below certain columns identify sites from one particular patient. Note that errors in different functions are evoked from sites in different zones in this patient. The individual maps of four other patients are given in Ojemann and Mateer (1979). Note also the presence of sites with changes in phoneme identification and orofacial mimicry (usually also with changes in naming or reading) in the posterior frontal, anterior parietal and anterior temporal zones, and the frequent specialized sites where only a single, language-related function is altered, in the surrounding anterior frontal, posterior parietal and posterior temporal zones. The 14 cases are a typical sample of adults (mean age 28.4 years, range 17–49) with medically intractable epilepsy. All are on anticonvulsant medications, most often phenytoin or carbamazepine; 13 had temporal lobe epileptic foci and surgical exposures; 1 frontal; 5 are male. Preoperative verbal IQ's ranged from 81 to 120 (mean 99.4). Mean number of stimulating sites 10.4 (range 4–16). Mean stimulating current 5.3 mA (range 3–8) between peaks of the biphasic pulses (Ojemann 1982)

Acknowledgment. I would like to thank Dr. Wolfram Ziegler for his valuable advice on neurophonetic aspects of this work.

References

Benda CE (1969) Down's syndrome, mongolism and management. Grune and Stratton, New York

Bless DM, Abbs JH (eds) (1983) Vocal fold physiology: contemporary research and clinical issues. College-Hill, San Diego

Clara M (1959) Das Nervensystem des Menschen. Barth, Leipzig

Hines M (1940) Movements elicited from precentral gyrus of adult chimpanzee by stimulation with sine wave currents. J Neurophysiol 3:442–466

Hirano M (1981) Clinical examination of voice. Springer, Berlin Heidelberg New York

Hobson JA, McCarley RW (1977) The brain as a dream state generator: an activation-synthesis hypothesis of the dream process. Am J Psychiat 134:1335–1348

Ingvar DH (1983) Serial aspects of language and speech related to prefrontal cortical activity. A selective review. Hum Neurobiol 2:177–189

Jürgens U (1976) Projections from the cortical larynx area in the squirrel monkey. Exp Brain Res 25:401–411

Jürgens U (1979a) Anatomical and functional cerebral organization of phonation in animals. Exp Brain Res Suppl II. Hearing mechanisms and speech. Springer, Berlin Heidelberg New York, pp 171–182

Jürgens U (1979b) Neural control of vocalization in nonhuman primates In: Steklis HD, Raleigh MJ (eds) Neurobiology of social communication in primates. Academic Press, London New York, pp 11–44

Jürgens U (1984) The efferent and afferent connections of the supplementary motor area. Brain Res 300:63–81

Jürgens U, Cramon D von (1982) On the role of the anterior cingulate cortex in phonation: a case report. Brain Lang 15:234–248

Jürgens U, Ploog D (1970) Cerebral representation of vocalization in the squirrel monkey. Exp Brain Res 10:532–554

Jürgens U, Ploog D (1976) Zur Evolution der Stimme. Arch Psychiat Nervenkr 222:117–137

Jürgens U, Ploog D (1981) On the neural control of mammalian vocalization. Trends Neuro Sci 4: 135–137

Jürgens U, Pratt R (1979) Role of the periaqueductal grey in vocal expression of emotion. Brain Res 167:367–378

Jürgens U, Hast M, Pratt R (1978) Effects of laryngeal nerve transection on squirrel monkey calls. J Comp Physiol 123:23–29

Jürgens U, Kirzinger A, Cramon D von (1982) The effects of deep reaching lesions in the cortical face area on phonation. A combined case report and experimental monkey study. Cortex 18: 125–140

Kirzinger A, Jürgens U (1982) Cortical lesion effects and vocalization in the squirrel monkey. Brain Res 233:299–315

Kohler KJ (1977) Einführung in die Phonetik des Deutschen. Schmidt, Berlin

Kuypers HGJM (1958) Corticobulbar connexions to the pons and lower brain-stem in man. Brain 81:364–388

Larsen B, Skinhöj E, Lassen NA (1978) Variation in regional cortical blood flow in the right and left hemispheres during automatic speech. Brain 101:193–209

Lassen NA, Larsen B (1980) Cortical activity in the left and right hemispheres during language-related brain functions. Phonetica 37:27–37

Lenneberg EH (1967) Biological foundations of language. Wiley, New York

Leyton ASF, Sherrington CS (1917) Observations on the excitable cortex of the chimpanzee, orang-utan, and gorilla. Q J Exp Physiol 11:135–222

Liberman AM, Cooper FS, Shankweiler DP, Studdert-Kennedy M (1967) Perception of the speech code. Psychol Rev 74:431–461

Lieberman P (1973) On the evolution of language: a unified view. Cognition 2:59–94
Lindblom B (1983) Economy of speech gestures. In: MacNeilage PF (ed) The production of speech. Springer, Berlin Heidelberg New York, pp 217–245
Maurus M, Kühlmorgen B, Wiesner E, Barclay D, Streit KM (1985) "Dialogues" between squirrel monkeys. Lang Commun 5:185–191
Morasch H, Cramon D von (1984) Laryngoskopische Befunde bei Dysphonie nach traumatischem Mittelhirnsyndrom. HNO 32:13–16
Ojemann GA (1982) Models of the brain organization for higher integrative functions derived with electrical stimulation techniques. Hum Neurobiol 1:243–249
Ojemann G, Mateer C (1979) Human language cortex: localization of memory, syntax and sequential motor-phoneme identification systems. Science 205:1401–1403
Penfield W, Roberts L (1959) Speech and brain mechanisms. Princeton Univ Press Princeton
Ploog D (1981) Neurobiology of primate audio-vocal behavior. Brain Res Rev 3:35–61
Ploog D (1985) Stimme und Sprechen unter der Kontrolle des Gehirns. In: Karlson P (ed) Information und Kommunikation, naturwissenschaftliche, medizinische und technische Aspekte. Verh Ges D Naturforsch Aerzte 113:pp 113–137
Robinson BW (1976) Limbic influences on human speech. Ann NY Acad Sci 280:761–771
Sutton D, Larson C, Lindemann RC (1974) Neocortical and limbic lesion effects on primate phonation. Brain Res 71:61–75
Sutton D, Trachy RE, Lindemann RC (1981) Monkey vocalization: effects of supplementary motor damage. Soc Neurosci Abstr 7:240
Vogel M, Cramon D von (1982) Dysphonia after traumatic midbrain damage. Folia Phoniatr 34: 150–159
Vogel M, Cramon D von (1983) Articulatory recovery after traumatic mutism. Folia Phoniatr 35: 294–319
Wilson EO (1975) Sociobiology. The new synthesis. Belknap Press of Harvard Univ Press, Cambridge, Mass
Winter P, Ploog D, Latta J (1966) Vocal repertoire of the squirrel monkey *(Saimiri sciureus)*, its analysis and significance. Exp Brain Res 1:359–384
Winter P, Handley P, Ploog D, Schott D (1973) Ontogeny of squirrel monkey calls under normal conditions and under acoustic isolation. Behaviour 47:230–239
Ziegler W, Cramon D von (1986) Spastic dysarthria after acquired brain injury: an acoustic study. Br J Disord Commun 21:173–187
Ziegler W, Cramon D von (1988) Zentrale Stimmstörungen. In: Springer L, Kattenbeck G (eds) Interdisziplinäre Reihe zur Theorie und Praxis der Logopädie, vol 3. In press

Glossary of Terms [1]

Definitions are to be taken as working definitions, not necessarily the most precise and technical definitions. They sometimes may be too brief or opinionated, but the authors run that risk for the benefit of keeping the glossary simple, clear, and helpful to the reader who is not so familiar with the subject.

ARTICULATION

Modifications of the signal produced by the glottal source as a result of alterations in the volume and configuration of the supraglottal tract (laryngeal air sacs, expanded hyoid bones, buccal cavity, nasal cavities, etc.) or from approach and contact between structures in the supraglottal tract.

ATONALITY

A signal is atonal, if it lacks the criteria of tonality. Most often, atonality is perceived by the human ear as noisiness.

ATTENUATION

Loss of sound pressure level, due to absorption, reflection, spherical sprading, etc., as the signal propagates through the atmosphere. Sound is attenuated by about 8 dB per doubling of distance from a point sound source [attenuation = $1/r^2$, r = distance ("inverse square law")]. See: Excess attenuation.

AUDIENCE EFFECT

The term audience effect describes the occasional dependence of communicative behavior on the availability of an audience.

BACKGROUND NOISE

The total of all sounds that effect the production and reception of biologically relevant signals. The Lombard Effect refers to the phenomenon that animals increase vocal

1 John G. Robinson, Jay M. Whitehead, Univ. of Florida, Dept. of Wildlife and Range Sciences, Gainesville, Florida 32611, USA and the editors

amplitude by 2 dB for every increase of 10 dB in background noise. The Cocktail Party Effect refers to the ability of animals to detect a biologically relevant sound from high levels of background noise.

CALL TYPE

Vocal patterns invariately appearing in the same uniform configuration of structural properties.

CATEGORICAL PERCEPTION

Usually based on behavioral responses, the receiver is considered to place vocal signals which vary along a structural continuum into discrete categories.

CONVERSATION

Conversation is given when the communicative process meets the following criteria: (1) evidence of turn taking; (2) directionality of change in acoustic structure; (3) bi-directionality of information transfer.

DOMINANT FREQUENCY

The dominant frequency is the spectral component with the highest amplitude.

EXCESS ATTENUATION

Attenuation in excess of that predicted under the inverse square law (e.g., resulting from sound-absorbing surfaces).

EXTERNAL REFERENT

Any attribute of the outside world which a vocal signal designates. This can be an external object, event, property, or relationship.
See: Internal referent

FFT

Fast Fourier Transform. A mathematical algorithm to transform data from the time domain to the frequency domain. The result of an FFT is the spectrum of the data points being transformed. The FFT is frequently used to compute series of spectra to form a spectrogram.

FORMANTS

Frequency bands with amplitudes modulated by supraglottal resonances. The first formant (the lowest) is called F_1, the second F_2, and so on. F_0 (the fundamental frequency), F_1 and F_2 (the formants) are among the most important spectral parameters. Note that F_0 describes a physically different effect than F_1 and F_2.

FREQUENCY DOMAIN

If a signal is measured as a function of frequency, it is obtained in the frequency domain. Data in the frequency domain are typically observed on a spectrograph.

FUNCTIONAL BOUNDARIES

Structural differences that mark functional differences.

FUNDAMENTAL FREQUENCY

The slowest frequency component of a harmonically oscillating system. The fundamental frequency (symbolized by F_0) is in almost all instances identical with the excitation frequency forcing the system to oscillate. The fundamental frequency is the frequency domain equivalent to the pitch period in the time domain.

GLOTTAL SOURCE

The anatomical structure, composed of the glottis and vibrating vocal folds, which produce the fundamental frequency (F_0) in mammals.

HARMONICS

Frequency bands that are integer multiples of the fundamental frequency. The fundamental frequency (F_0) is said to be the first harmonic. The frequency of the second harmonic is twice the F_0, the third harmonic is three times the F_0, etc.

INTERNAL REFERENT

Any attribute of an internal variable which a vocal signal designates.

JITTER

Frequency perturbation that reflects erratic vibratory patterns of the glottal source.

LEXICAL SYNTAX

A syntax generating vocal sequences, the elements of which are related to a specific context, and have specified functions and behavioral consequences. This syntax involves the use of vocal sequences which derive their meaning from the multiplexing of the meaning of the component elements. Lexical syntax in human linguistics is the set of rules that account for and predict the combination of meaningful sounds (= morphemes or words) into meaningful sequences. Lexical syntax includes both lexical rules, which generate the lexicon, dictionary, or list of words, and grammatical rules, which generate sentences.

MASKING

The process by which the detection of signals is impeded by the presence of a (similar) sound.

OVERTONE

Frequency bands that are multiples of the fundamental frequency, although not necessarily integer multiples. Bells, for instance, have many nonharmonic overtones.

PHONOLOGICAL SYNTAX

A syntax generating vocal sequences, the elements of which are not related to a specific context, neither do they have specified functions or behavioral consequences. This syntax involves the use of vocal sequences whose meaning is unique to entire sequences. Phonological syntax in human linguistics is the set of rules that account for and predict the combination of meaningless sounds (= phonemes) into meaningful sequences (= morphemes or words).

PITCH

Traditionally, the subjective correlate of frequency. The term pitch is often (inaccurately) taken as a synonym for fundamental frequency.

PITCH CONTOUR

The time course of the fundamental frequency in a continuously tonal signal. Sometimes referred to as the pitch profile or F_0-contour.

PITCH EXTRACTION

The automated determination of the fundamental frequency in a spectrum.

PITCH PERIOD

The duration of a full period of the fundamental frequency in the time domain.

SPECTRUM

A distribution of amplitude across frequencies in a specified interval of time.

RESONANCE

The alteration (both damping and amplification) by the supraglottal structures of the signal produced by the glottal source. The resonant frequencies produced need not be harmonically related to the fundamental, see Formants and Harmonics. Supraglottal (= resonant) structures can also suppress specified frequency bands produced by the glottal source (anti-resonance), including the fundamental frequency.

SEMANTICS

Generally the study of the "meaning" of sound signals, more specifically the relationship between sound signals and their referents.

SYNTACTICS

The study of order in a sequence or series of sound signals.

SYNTAX

A system of rules that generates a sequence of vocal elements. This sequence occurs in a specified context and results in a predictable behavioral response. Syntax is indicated when vocal sequences are ordered and predictable. See Phonological Syntax and Lexical Syntax.

TIME DOMAIN

If a signal is measured as a function of time, it is measured in the time domain. Data in the time domain are typically observed on an oscilloscope.

TONALITY

Characteristic of vocalizations with a fundamental frequency and/or corresponding harmonics. A tonal vocalization is the result of the sound source periodically oscillating.

ZOOSEMIOTICS

The discipline devoted to the scientific study of signaling behavior in and across animal species.

Subject Index

MIX
Papier aus verantwortungsvollen Quellen
Paper from responsible sources
FSC® C105338

If you have any concerns about our products,
you can contact us on
ProductSafety@springernature.com

In case Publisher is established outside the EU,
the EU authorized representative is:
Springer Nature Customer Service Center GmbH
Europaplatz 3, 69115 Heidelberg, Germany

Printed by Libri Plureos GmbH
in Hamburg, Germany